# Man-Eaters

# Man-Eaters

## Michael Bright

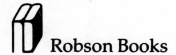

Robson Books

Published in 2000 by Robson Books, 10 Blenheim Court, Brewery Road, London N7 9NT.

A member of the Chrysalis Group plc

British Library Cataloguing in Publication Data:
A catalogue record for this title is available from the British Library.

ISBN 1 86105 321 5

**Printed by Redwood Books, Trowbridge, Wiltshire**

# CONTENTS

# CHAPTER 1

---

# WHY MAN-EATERS?

W e shudder understandably at the thought of being eaten alive, but it is not a new feeling. About 30 million years ago, our earliest recognisable ancestors were clambering about in the trees in what are now the deserts of north-east Africa. Their fossilised remains were discovered at Fayum in Egypt by Daniel Gebo and Elwyn Simons, of Duke University in North Carolina, but when the two anthropologists came to examine the skulls they found puncture marks that had been made by the canine teeth of creodonts (primitive mammalian carnivores that are now extinct). While our earliest primate anthropoid ancestors were plucking the odd insect from the branches, they themselves were being snatched from the trees by wily predators.

Many millions of years later, came a time when another group of early ancestors dropped down from the trees and were exposed to the dangers on Africa's grassy plains. But as we were on the way down, a significant predator was on the way up. It was the leopard, probably the most successful of the big cats and a known man-eater. It was consuming our early ancestors and their close relatives from the moment they hit the ground.

1

Robert Brain, director of the Transvaal Museum in Pretoria, has been studying robust ape-man fossils (*Australopithecus*) that were found in the limestone caves at Swartkrans in South Africa. These heavily built hominids were ancient plant-eating, gorilla-like apes that lived at the same time as our direct ancestors. He examined their skulls and found they had distinctive perforations. Some of the holes were clearly made by sabre-toothed cats, while others were ¼in (6mm) across and found in pairs, each hole about 1¼in (32mm) from its neighbour. At first, it was thought that blows from a pointed weapon were responsible, but more recent work has shown that they resemble the tell-tale puncture holes made by the large canine teeth in the upper jaws of big cats, in particular, those of leopards.

He compared the cave remains with the debris left behind by modern carnivores. The holes in the skulls matched exactly teeth marks in the skulls of baboons killed by leopards. Baboons regularly fall prey to leopards, especially at night in their sleeping sites. The leopards ambush them, just as they grabbed those ancient man-apes asleep in their caves 1.6 million years ago.

Remarkably, the Swartkrans caves have very few skulls of our direct ancestors (*Homo*). It seems, even one-and-a-half million years ago, we were already clever enough not to sleep where leopards were likely to attack. Indeed, at this stage in our prehistory we began to beat the large predators at their own game. When we turned to eating meat as a substantial part of our diet, we probably scavenged first on the kills of the big cats and hyenas and then went hunting ourselves.

By living together, co-operating and communicating, we became effective hunters. We also developed the wherewithal to be effective competitors. Hunting technology – first stones, and then slingstone, spear, bow-and-arrow, cross-bow, blow-pipe, harpoon, bolas, boomerang, throwstick, rifle, shotgun and pistol – gave us the upper hand, but it has never removed the fear. Strip away our technology and the rest of our troop, and a person alone, becomes relatively weak and defenceless. In short, individually we are fair game for powerful predators, and in parts of the world where people rub shoulders with lions, tigers, crocodiles and sharks, it is we who are sometimes the prey.

Despite this conspicuous vulnerability, wild animals – even big and powerful ones – are curiously wary of people. Under normal circumstances, they avoid us. Attack, of course, is a successful strategy to adopt in defence. Many people have been attacked by cantankerous buffalo or elephants that just do not like us to be in their company, mother bears protecting their youngsters, or venomous snakes that have been surprised and lash out in self-defence. They might kill us but they do not eat us. Many are plant-eaters and have no dietary interest in people anyway, but on occasions some meat-eaters actually see us as food. So, what makes them turn to man-eating?

The most commonly accepted reason is simply 'old-age'. Old animals with broken teeth, arthritic limbs, and worn claws settle for an easy life. They take anything that requires the minimum of effort to catch. People are easy meat. We can't run fast enough to get away, even from a lame predator. We are easily taken by surprise and without weapons we cannot defend ourselves. But is that all there is to it? Healthy animals take to man-eating too, and almost everywhere in the world today, the reason is simple and clear to see. Predators are losing their living space. It is disappearing at an alarming rate and with it their natural prey. In some areas poaching has presented predators with direct competition. Guns, traps, snares and nooses take out game, and predators go hungry. Humans are a convenient alternative.

Roads, railways and expanding villages and towns mean that animals come into contact with people more often. Our wild neighbours lose their natural suspicion of humans, and triggered by hunger, go on a man-eating spree. And, having learned the ease with which we can be caught, adult predators pass the skill on to their offspring. Young predatory mammals, such as lions, tigers and leopards, might learn from their mothers that people are good to eat and easy to catch.

Some predators are introduced to man-eating after wars and natural disasters. At first corpses are scavenged but, when these have been devoured, living people are a convenient alternative. This happened in the Arakan district of Burma during the Second World War. Tigers scavenged the dead bodies of soldiers who were left behind during the retreat of 1942. Then, they took to attacking and eating the living.

There are also cases of mistaken identity, particularly if we find ourselves in the wrong place at the wrong time. Great white sharks, for example, may confuse people on surfboards for seals, and tiger sharks see people on body-boards as sea turtles. Ignorance in these situations is not bliss. Surfing near a seal colony in seas frequented by great white sharks or near a beach where turtles haul out to deposit their eggs is simply asking for trouble.

Indeed, careless behaviour by people is more than often the cause of man-eating incidents rather than aberrant behaviour on the part of wild animals. Take the case of three men who camped in the Serengeti National Park. Their tent was so small, their heads protruded from one end. Two inquisitive lions passed by, spotted the row of heads and went to investigate. One of the men woke up and, seeing the lion so close, screamed. The lion did what came naturally – it bit. The man died before he could be flown to hospital. His head had been crushed and his skull broken.

Some cases of man-eating are not that at all. Sometimes a person might stumble on a tiger or leopard with cubs, and the mother will naturally attack in defence of her young. She might take advantage of the unexpected availability of some fresh meat but she is not necessarily a compulsive man-eater, even though often as not she is labelled as such.

Then, of course, not all so-called man-eater attacks are due to wild animals. In Africa, secret societies of lion-men and leopard-men (the Aniotos) and, in Europe sects of human werewolves, were actually contract murderers or 'hit-men'. They wore animal skins and carried iron claws, and caused wounds on their victims that resembled attacks by wild animals. They settled old scores – for a price – and innocent animals were killed in revenge.

People, especially debtors, cheats and fraudsters, also conveniently disappear, the so-called victims of wild animal attacks. Strange, though, that they should turn up perfectly well many months or even years later. In the Bastar area of India, a woman was alleged to have been killed by a man-eating tiger but when police investigated the case fully they discovered that her husband was the murderer. He had wanted to marry another woman. In the meantime, the tigers of Bastar and their cubs were slaughtered – all declared 'man-eaters' before the truth was revealed.

In the wild, it is surprising perhaps that people are not attacked or killed more often. Given the number of people at any one time who bathe, swim, surf, wind-surf, kayak or immerse themselves in the world's seas, for instance, the shark-attack statistics should be far higher, that is, if sharks had a predilection for human flesh. But they do not. Sharks, like most predators, do not seek us out.

Compared to a big cat, bear or large shark, however, we are relatively weak and vulnerable. Our advantage is the ability to co-operate, communicate and look into the future or past, learning by our mistakes and planning to put things right. In effect, our 'intelligence' keeps us out of trouble. We also have the technological wherewithal to eliminate man-eaters. In this way, any genetic propensity to feed on people is less likely to be passed on to subsequent generations.

Nevertheless, there is an underlying fascination and fear for animals that could eat us, particularly those that actually do so occasionally. One way in which we deal with this fear of the wild is simply to eliminate it in what amounts to a ritualised display of superiority. Down the ages, people slaughtered millions of lions, leopards, tigers and anything else that seemed remotely dangerous. The 'bag' was used to inflate egos, and provide dinner-table tales for intrepid hunters. But, most of all it dealt with that fear.

The result has been that man-eaters have more to fear from man, than man has to fear from man-eaters. Since ancient times, people hunted the larger predators for sport. It was a pastime reserved for kings and other lesser nobles. Pharaoh Amenophis III, who reigned from 1405 to 1367BC, is reported to have slain 102 lions during his first ten years on the throne. Several centuries later, Ashurbanipal of Assyria split the heads of lions with his two-handed sword. At the time lions were breeding well, so well, in fact, they were running out of natural prey and turning to stock-killing and man-eating.

Lions – at least dead ones – also pleased the Roman general and dictator Julius Caesar in the first century BC. He had 400 killed ceremonially to consecrate the Forum, and when Germanicus Caesar was nominated as consul some years later, another 200 were ritually slaughtered. Brown bears and polar bears were also placed in the arena where they were pitted against dogs and gladiators. In one day in AD237 a Roman noble who later became Emperor Gordian I was responsible for a 'contest' in the Coliseum during which 1,000 bears were said to have been despatched by gladiators. The event successfully exterminated local bear populations and subsequently bears had to be imported from northern Europe and North Africa.

During the 19th century in California, bears were placed in arenas with bulls. The practice was thought to have been responsible for the demise of the California grizzly. Bears, lions and tigers were also to star in 'baiting' contests, a practice that continued until the mid-19th century in Europe. The captured predator was beaten, blinded and chained to a stake in the baiting pit where packs of dogs and men with whips goaded and harassed it until it was exhausted and killed.

The tiger was the unwitting plaything of Indian Maharajahs and British colonels. From 1821–8, no less than 1,053 tigers were shot in the Bombay Presidency. In the 1850s Colonel William Rice shot 93 tigers in four years, but the Indian princes notched up far more: during the 1920s, the Maharajah of Kotah shot 334 tigers, while the Maharajah of Rewa beat him with a tally of 364. The Maharajah of Udaipur almost made it to the top with a bag of 1,000 kills, but the top tiger killer was the Maharajah of Surguja who died as recently as 1958. Before he expired, however, he accounted for 1,707 tigers.

# CHAPTER 2

---

# WOLVES AND WILD DOGS

Since time immemorial people in northern lands have had a deep inner fear of being attacked and eaten alive by wolves (*Canis lupus*), and it is reflected even today in the way we speak. In the English language, we have expressions such as 'wolf at the door', 'wolf in sheep's clothing', and 'to be thrown to the wolves'. The word 'wolf' became synonymous with evil, and often it has been associated with the villains in history. During the Second World War, Hitler's forest headquarters in East Prussia was the 'Wolf's Lair' and flotillas of Nazi submarines were known as 'wolf-packs'. Further back in history, Isabella of France, consort of Edward II had her husband murdered by the thrust of a red-hot poker into his bowels. The act earned her the title 'She-wolf of France'.

The word 'wolf' has crept into all manner of activities. In music, a dissonance in some chords is known as the 'wolf tone', in biology the voracious ground-hunting spider is a wolf spider, and in marine biology there are fish with ferociously big teeth known as wolf-eels, wolf-fish and wolf-herrings. Anyone cruel or lustful is said to be wolfish, and a diner might wolf down his or her food.

This deep fear and revulsion of the wolf led to myths, fairy tales and legends. The story of *Little Red Riding Hood* is probably the most commonly told, although the tale of the *Three Little Pigs*, immortalised in the song 'Who's afraid of the big bad wolf?', is the latest in a long line of children's stories to perpetuate the image of the wolf as a miscreant.

Even in these enlightened times the presence of a wolf in some parts of Scandinavia and the USA triggers the most extraordinary bouts of mass hysteria. It is as if society is exorcising some ancient demon. In due course, newspapers and television carry pictures of jubilant hunters standing over the dead animal. The inevitable consequence is that in many places, including the British Isles, the wolf has been persecuted to extinction.

Wolves, however, are not supposed to attack and devour people; at least, that's what a wolf conservationist will tell you. Scientists have looked long and hard at claims down the centuries, but still the official line is that there have been no substantiated or authenticated cases of wolves making unprovoked attacks on humans. Yet the history books and newspapers are full of accounts of wolves up to no good, and in parts of India, where village folk are subject to harassment from all manner of beasts, there seems to be a fatal attack almost daily. Whether these reports contain elements of truth or whether they are simply instances of mistaken identity, acts of cold-blooded murder, people mysteriously disappearing to escape paying old debts, or figments of a vivid imagination, we never seem to get to the bottom of them, but they certainly make harrowing reading.

The wolf appears to have earned its reputation as a people-killer some time after the fall of the Roman Empire. Wolves took livestock mainly, but according to ancient accounts, they killed and ate people too. Indeed, during the time of the Emperor Charlemagne (742–814), the populace was so worried about wolves that it established the first known formal wolf hunts. Wolf hounds were specially bred for the purpose and professional wolf hunters were employed to rid the countryside of its wolf population.

Wolves across Europe increased in numbers at this time because the forests in which they lived were recovering from the ravages and excesses of the Roman occupation. An increase in forest cover furnished wolves with more living space. But, in the early part of the Middle Ages, around AD1000, a renaissance in agriculture, with the introduction of the heavy plough and the three-field farming system, meant that the forests were cut down once more. Wolves inevitably came into close contact with the rapidly increasing human population. Livestock was taken and, if the reports are to be believed, so were people.

In many cases, the wolves were simply exploiting a source of readily available carrion. During the devastating pan-European epidemics of plague, such as the Black Death between 1347 and 1351, and the many wars of these times, human corpses must have littered the uninhabited country-side. They would have been left to rot – food for wolves and other scavengers. Maybe the wolves even dug up and ate recently dead bodies. It is, perhaps, significant that the Latinised Germanic word for a grave robber is *wargus*, the root word *varg* meaning wolf in northern European languages, such as Swedish. It dates from about AD500.

The problem for modern scholars who attempted to distinguish fact from fiction, however, was that the supernatural dominated popular thinking in the Middle Ages, and accounts at the time were more than tainted with a spoonful of myth and mysticism. Wolves were thought to be at one with the devil. It was the time of the werewolf, and other malevolent creatures of the night. Extracting the truth is understandably difficult, but the reports of people being killed and eaten by wolves are far too numerous to ignore.

St Francis of Assisi (1182–1226) was preaching at the town in the north of Italy when his attention was brought to the infamous 'Wolf of Gubbio'. It had been terrorising the town's inhabitants. They were terrified to leave the city for fear of being attacked. The legend tells that St Francis confronted the wolf, which tried to attack him, and persuaded it to give up its man-eating ways in exchange for being fed regularly by the townspeople.

One of the classic wolf stories is that of Courtaud, a wolf that lived in 15th-century France. It was the alpha male of a pack of a dozen or so wolves and easily recognised by its bobtail. It gained notoriety during a killing spree in the summer of 1447. At the time, Paris was a walled city centred on a single island – the Île de la Cité – in the River Seine, and livestock was driven regularly along tracks leading to the city gates. Lying in ambush was Courtaud and his pack. They not only intercepted the livestock, but also killed the people who were herding them. On one occasion, in February 1450, the pack entered Paris itself through a breach in the city walls and killed 40 Parisians – or so the story goes.

The wolves lived and bred in caves on the north bank outside the city walls, a place known as Le Louvrier – the place of the wolves – and now the site of the famous art gallery Le Louvre. The city dwellers tried to eliminate them in their lairs and failed; the wolves were too clever for them. Eventually, Courtaud's pack was enticed with a trail of meat into the square in front of Notre Dame. The exits were blocked and every wolf was stoned or stabbed.

Although the wolf had been eliminated from England by the beginning of the 16th century, it was still common in Scotland for another century. As natural game was hunted out, wolves turned to sheep and cattle for food, and sometimes to people. As late as the 1740s, there are stories of a large black wolf that was despatched eventually by MacQueen of Pall-a-chrocain. The animal had allegedly killed two children who were crossing the moors with their mother at Findhorn. Such was the fear of wolves previous to this that 'spitals' or rest houses were built to provide travellers with havens safe from marauding wolf packs. Dead people were not buried in remote churchyards for fear that they would be exhumed and consumed by wolves. Wolf hunts were ordered, particularly when cubs were being reared in early summer. During James IV's time, all men of military age were forced to take part in wolf hunts at least three times each year. The black wolf was the last wolf in Scotland to be killed. The year was 1743, although some believe that the story is a fairy tale and that Scottish wolves disappeared some years earlier.

Somewhat later, between 1764 and 1767, another rascal was described by the Abbé François Fabre. This wolf was labelled 'The Beast of Gevaudan', and it frequented the Gevaudan region of southern France. It was a big male, estimated to be about 130lb (59kg) that ran with a large 109lb (49kg) female. Adult wolves average 60–120lb (27–54kg) today, European specimens tending to be smaller than their North American counterparts. But the Gevaudan wolf was to all accounts a formidable beast which reputedly attacked more than 100 people, killing and eating at least parts of 64 of them. It has been suggested by Dr C. H. D. Clarke of Ontario, who has made a study of these animals, that the beasts of Gevaudan were not pure wolves at all, but dog-wolf hybrids.

Other distinguished scientists and public figures have also contributed to the debate. In 1890, St George Jackson Mivart (an English biologist and one of Darwin's critics) chronicled 161 human deaths from wolf attacks in Russia during 1875. There were even stories of packs of a hundred or more wolves attacking and killing hunters who had been scouring the forests in attempts to track them down. And, during the campaign at British-held Jalalabad in Afganistan, between 1878 and 1879, General Kinloch reported that the sentries guarding his camp were often attacked by wolves during the night.

In northern Europe, where the wolf has always enjoyed a sinister reputation, there were two years – 1880 and 1881 – which were called the 'wolf years' in Finland. Wolves, so it is said, killed 22 children aged between two and nine years old in an area around the town of Turku (Abo) in the south-west corner of the country. One story tells of the eight-year-old son of D. Hornberg, a tailor, being taken from the family farm near Nykyrko by a large wolf. The animal is said to have carried the boy in its mouth – no mean feat in itself – to the forest and eaten him.

More recently, in October 1995, a 33-year-old man was attacked by a wolf at Seipajarvi – about 560 miles (900km) north of Helsinki. The wolf was more interested in the refuse cans at the back of the house, but when the man went to investigate the clanging noise, the wolf attacked him viciously. The two wrestled for about ten minutes, and then the enterprising Finn got his own back – he bit the wolf.

Interestingly, there are wolf attack stories emanating from different parts of Europe that are remarkably similar. There is, for example, the story of a soldier, or maybe a postman or farmer, camping with his dog, depending on the source of the tale, who is out in the forest on a bitterly cold winter's night when he is attacked by wolves. He pulls out his sword and puts up a spirited fight, wounding some of the pack. The wolves withdraw and he sheathes his sword. But the pack returns and attacks again. In the meantime, the blood on the sword has frozen and the man is unable to pull out his sword a second time. Only a few pieces of skin and bones are left on the ground together with the ill-fated sword, the only evidence that a tragic event has taken place.

Clearly the same tales have done the rounds, and maybe the original story held some grain of truth. It is probable that people in ancient times did encounter wolves and some undoubtedly became wolf food. At times when natural prey was considerably reduced and farm animals were scarce, there is every likelihood that man was on the wolf's menu, and conditions like these may well have occurred during the Middle Ages, a time when many of these man-eating stories most probably arose.

Similar conditions sometimes occur in more modern times. Russia, for example, has had its fair share of wolf tales. In 1927, for example, a pack of wolves was said to have besieged a village, picking off the inhabitants one by one until the armed forces came to rescue the survivors. And in the 1960s reports came out of the Ural Mountains that described wolves attacking 168 people and devouring 11 of them.

In this part of the world, fluctuations in wolf populations have been closely linked to fortunes of the human population. During the hard times after wars or economic difficulties, people cared more about their own survival than to worry about wolves. Consequently, wolf populations increased dramatically. In 1940, for example, there was an estimated 200,000 wolves in Russia, and their impact on farm livestock was devastating. Wolves were reduced in number dramatically by about 50,000 in a single year. Then again in the early 1970s numbers rose, and in the mid-1980s wolves increased their population sixfold, with over 100,000 present in the summer of 1984. Wolves even wandered into housing estates in the pine and birch forests on the outskirts of Moscow and other cities. In the countryside, over 50,000 sheep, cattle and horses were estimated to be killed by wolves each year, and wolf attack stories on people began to appear in the newspapers. Hunting was encouraged again, sometimes with helicopter gunships and automatic weapons. Unfortunately (or fortunately, depending on whose side you are on) the wolves were too clever: they began to recognise the sounds made by the hunters' helicopters and aerosledges and hid.

The Russian studies showed, however, that the wolves were important in keeping populations of other animals healthy. When wolves were hunted heavily, the number of sick caribou increased on the Taimyr, the Darwin elk herd was reduced by serious skin disease, and there was a recognisable degeneration of red deer in the Crimea. Wolves target their prey and eliminate sick and weak animals.

According to more recent reports picked up by the BBC Monitoring Service, wolves have also been reducing livestock and harassing people in Kyrgyzstan. In January 1999, wolves attacked communities in the two largest regions – Naryn and Issyk-Kul – in the south-east part of the country. The wolves make daily incursions on livestock and have attacked people, although there have been no deaths to date. Local administrators stated that people were afraid to go out of doors.

Turkey has also had a number of wolf incidents. Again, the absence of natural prey seems to have triggered the attacks. On 14 January 1968, wolves were reputed to have killed and eaten two villagers during a period of severe cold, when over 6ft (2m) of snow covered the ground in the province of Bolu, to the northwest of Ankara. And, in November 1983, a major earthquake hit the eastern provinces of Turkey and many refugees were sheltered from the bitter snow storms in makeshift, tented camps. Wolves approached the camps in search of food, and soldiers fired into the night to scare them away. Whether the people were really in danger is anyone's guess.

A previous incident indicates a possible explanation for some wolf attacks. A newspaper article from Associated Press in 1962, describing an attack on villagers of Kaynaklar in the Buca district of Turkey by twenty or more wolves, suggested that the raid was probably the result of an epidemic of rabies. A nine-year-old boy was killed and 14 other people injured as the wolves were repelled with axes, scythes and just about any implement the farmers could lay their hands on. The assault on the village went on for about seven hours.

Undoubtedly, some genuine wolf attacks have been the result of rabies – the fearsome disease that causes dogs and wolves to foam at the mouth and literally to go berserk. In France, between 1851 and 1877, records show that 38 people died from rabies after having been bitten by wolves (during the same period a further 707 died after bites from dogs). And in 1942, a man riding a railroad 'speeder' (a small flat-topped railway wagon propelled by pumping a handle up and down) in Canada was attacked repeatedly for about a half-an-hour by a wolf. The man used an axe to defend himself and several people came to help until the beast was killed. The animal's tenacity led authorities at the time to conclude that it was rabid.

Other explanations are less biological. Wolf enthusiast Erik Zimen recalls a time in September 1975 when the German Press Agency released a story about a pack of wolves attacking a village in Italy's Abruzzi Mountains, an area where Zimen and Italian wolf expert Professor Luigi Boitani of Rome University had been studying the local wolf population. They went to investigate. It turned out that all the fuss was about whether a ski resort should be built or not. The local villagers were in favour but conservationists were not. Somehow, wolves were brought into the story because the local farmers had had problems with wolves attacking their sheep that year, and the protection of wolves was associated with the conservationists. One farmer alleged that there were over 50 wolves in the surrounding mountains and that it was not safe for children to go out alone. By the time the story was written up by a journalist in Rome it was completely out of hand and the headlines proclaimed that 50 wolves had besieged a village.

Similarly, Paul Joslin, of the Chicago Zoological Society, investigated claims of wolf attacks in the mountains of north-western Iran. Teheran newspapers apparently feature news of many attacks every winter, but Joslin found that most reports were of villagers travelling alone who chanced upon wolf packs and fled, believing they had diced with death but lived to tell the tale. Joslin looked into one report of a shepherd having died, and discovered that, together with another shepherd and some villagers, he had been successful in chasing away the wolves threatening his sheep, but then sat down, coughed and dropped dead.

Currently, wolf-attack mania is beginning to appear in the French Alps where no humans have been harmed but an estimated population of 300 wolves has been hitting flocks of sheep. One shepherd brought his sheep down from their 1999 summer pastures because 180 out of a 1,200-strong flock had been killed by wolves. Stories of attacks on people are yet to appear.

Curiously, most wolf-attack stories emanate from Europe and Asia. The wolves of North America appear to be a more docile bunch. Indeed, newspapers in North America once offered rewards for evidence to substantiate alleged wolf attacks, but the money was never claimed. The reality is that the European settlers in North America were better armed and provisioned than the peasants of central Europe in the Middle Ages. People with firearms could defend themselves, and the wily wolf would have learned quickly to stay well out of pistol range. They also had plenty of wild game to eat – bison, pronghorns, elk, and moose – and so humans would have been way down the list of preferred prey.

Nevertheless, North American wolves have been known to eat people. In the same way that the wars and pestilence in Europe provided Old World wolves with ample food supplies, similar gluts were served up to their New World relatives. In the mid-19th century, when the 'gold rush' saw many adventurers heading for the west coast of America, yellow fever took a considerable toll and the bodies were reportedly devoured by wolves. Moreover, American Indian tribes ravaged by smallpox may have attracted scavengers, including wolves.

15

There are, in fact, a few wolf-attack stories from North America. Some tell of farmers having been bitten when they have tried to separate two animals – the farm dog and a wild wolf – but who is to say that it was not the dog rather than the wolf that inflicted the injury? In October 1983, there was the case of a pet wolf that, together with two cubs, escaped from its pen at a farm at Malad, Idaho. It killed a three-year-old boy by biting him on the head and neck.

Although there are no substantiated cases of unprovoked attacks in the wild, records exist of the remains of people being found in bloodstained snow surrounded by the unmistakable tracks of wolves. There is no question that wolves would scavenge a human corpse, but had they been the slayers or had they simply chanced upon an easy meal? We shall never know. Suffice to say that wolf researcher Douglas Pimlott was once walking in a Canadian forest and felt that a pack was watching his every move. Indeed, he saw that one of its members was following a parallel course to his own. Understandably he felt very uneasy. Was his anxiety unfounded or was he in real danger?

Pimlott's apprehension may have some justification when we read of a group of Canadian researchers who encountered a wolf pack on 29 June 1984. The three scientists from the University of Toronto – Peter Scott, Catherine Bentley and Jeffrey Warren – were walking in spruce-lichen woodland to the south of East Twin Lake, about 9 miles (15km) south-east of Churchill, Manitoba. While they trekked, they had noticed a criss-cross of wolf tracks and scats. They entered an oval clearing and sat down to rest. Suddenly, they heard a twig snap and on turning round saw a cream-coloured wolf heading straight towards them. Bentley yelled and stamped her feet, and the wolf tried to stop and turn at the same time. It lost its balance and tumbled into a bush not far from the startled researcher. Regaining its balance, the wolf retreated, only to be replaced by a second black-mantled wolf which made a lunge at Scott, who was trying to extract a 'bear horn' (loud boat horn used to frighten away polar bears) from his pack. The animal leapt forward in 6ft (2m) bounds, ears erect, tail flat out and eyes locked on to Scott's. Scott sounded the horn and the wolf, momentarily surprised, lunged obliquely to one side. The researchers shinned up the nearest trees. For the next four hours, three or four wolves were spotted about 30 times. One came to the edge of the clearing and leaped about, barking, howling and whimpering. When no wolf had been spotted for 15 minutes, the group came down from the trees, formed a defensive back-to-back triangle and moved slowly, from tree to tree, back to their vehicle. On the way, they chanced upon a wolf den, and surmised that they had unwittingly walked into the wolf pack's rendezvous site nearby. Rather than seeing the researchers as an easy meal, more probably the wolves were concerned that these strangers had invaded their patch and were determined to see them off.

What the researchers had observed, though, is that wolves working together are formidable predators. What makes wolves such successful hunters is their ability to co-operate. A pack may consist of an alpha male and female – the breeding pair – together with several other less dominant adults, sub-adults and cubs. They communicate with body language, chemicals and sound, and this enables them to co-ordinate their hunts. The wolf howl is just one of their signals. It is used as a roll-call before the hunt to ensure every member of the pack is ready and willing to hunt, and again at the kill to bring in the flankers and stragglers which may have become separated from the rest of the pack. It also warns neighbouring packs not to intrude into the resident's territory. By co-operating in this way, wolves not only track and chase prey over long distances, but also bring down animals far larger than themselves.

Whether wolves see people as food at all may depend somewhat on what other food is available. During hard times, when a wolf's normal food – anything from a creature as small as a mouse to one as big as a deer – is scarce, there is the possibility that it turns to prey closer to home. It becomes bolder, hunger overcoming its natural reluctance to be in the vicinity of people. Farm livestock is particularly vulnerable to attack. It is easy prey, either confined in a field or bunched in a free-ranging herd, and there are numerous reports of wolves having a field day with a rancher's stock.

There was, for example, the notorious 'Custer Wolf', named after the town of Custer, South Dakota. It is reputed to have accounted for about $25,000 worth of cattle while hunting the Wyoming-Dakota border between 1910 and 1920. The wolf – an easily recognised white wolf – apparently killed for pleasure, despatching 30 sheep on one night, while eating only part of one, and killing 10 steers on another. In reality, the wolf did not gain any enjoyment from the slaughter. Presented with a super-stimulus and victims unable to escape, it did what comes naturally – it killed anything and everything within killing range, like a fox in a chicken coop. It was simply responding to an innate urge to maximise its store of food when presented with a ready available bonanza. More than likely, if left to itself the wolf would have returned to the killing site and stored the food about its territory for leaner times, a behaviour known as scatter hoarding. Nor was it alone. According to eyewitness reports, two escorts – a pair of coyotes – were in attendance throughout its infamous career. They scavenged on the carcasses abducted by the wolf that is, until they were confronted by Wyoming's ace wolf hunter H. P. Williams who shot them dead. Williams had recently shot another notorious wolf, the Split Rock Wolf of Wyoming, and he also trapped and killed the Custer Wolf (after the animal had lunged at him, pulled the stake from the ground and limped for three miles or more with the trap clamped on to its leg) ending a decade of stock killing.

In more recent times, wolves in Canada and Minnesota have been known to attack people, but the incidents are very rare. Events as far apart as 1915 and 1970 saw wolves attacking first dogs and then the people who were trying to rescue them. Clearly the wolves were not targeting the people as food. And in 1982 near Duluth, a hunter was clawed but not bitten by a wolf. It was probably a case of mistaken identity; the hunter's clothes reeked of the smell of deer. One of the other cases was that of a young girl in Ontario in 1987. She shone a torchlight into a wolf's eyes and was bitten. All these attacks indicate defensive or threatening behaviour (or a mistake in one case), and not feeding. If these wolves had seen the people as prey, the outcome would have been quite different.

There was, however, a fatal attack on 18 April 1996 at the Haliburton Forest and Wildlife Reserve in Ontario, Canada. The victim was a 24-year-old woman who had been hired to look after the wolves in a new Wolf Centre at the reserve. The animals themselves were captive wolves, although they had not been socialised to humans. They lived wild in their woodland enclosure, and usually kept clear of any humans that entered. On this occasion, five wolves attacked the young woman. Her body was found and the police informed. When two officers tried to recover her body, the wolves growled at them, as if defending a food cache, and then circled them. The policemen were understandably nervous, fired shots close to the wolves and then backed out. Later six officers entered and removed the body. The woman's clothes had been ripped off, and she had pieces of flesh torn away, as well as multiple bite wounds on her body. The local coroner ordered the wolves to be put down.

Eric Klinghammer, Director of the Institute of Ethology at Nawpf-Wolf Park, Battle Ground, Indiana, USA went to investigate and after interviewing people on the spot, he worked out a likely scenario. The woman had had concerns about the alpha male wolf, but had only shared them with her fiancé rather than the reserve officials. Being relatively new to the job, she probably entered the enclosure to familiarise herself more with the place and its occupants. Fallen trees and branches littered the ground, so she might have tripped and fallen. This would have been all the wolves needed to attack.

In September 1998, another attack occurred in Canada, this time in Algonquin Park. A wolf grabbed a 19-month-old boy around the rib cage, and tossed him about 3ft (one metre) away. The baby needed stitches but survived. The wolf had been seen around the campgrounds during the summer. It had rummaged through backpacks and tried to drag away a sleeping-bag in which a child was sleeping. It also attacked three dogs. Later, it was hunted down and killed. It was the fifth wolf-biting attack in the park in 12 years.

Yet, while deliberate attacks on people in most parts of the world are yet to be accepted as reality by the scientific community, the situation in India appears to be different. India is not only home to the man-eating tiger and leopard, but also the man-eating wolf. During recent times, the local subspecies of grey wolf – the Indian grey wolf or Asiatic wolf (*Canis lupus pallipes*) – has been recognised as the perpetrator of attacks on people, particularly children. It hunts in pairs or small family parties.

Between October 1980 and March 1981, a small group of wolves worked a territory 19 miles (30km) long by 16 miles (25km) wide on the Andhra Pradesh side of its border with Karnataka in the south-west part of India. The prey included humans. The pack – an adult male and two adult females, two sub-adults and three cubs – accounted for the deaths of nine children, and a further twelve were injured. It always operated in the same way, hiding in bush or crops near villages and taking children who were walking alone, usually at dawn or dusk. Children accompanied by adults were not approached. The victim was attacked from behind, pushed to the ground and bitten on the head or neck. The skull was then opened and the brain consumed first. If the alarm was not raised and rescuers failed to appear, there was time enough for the wolves to eat the rest of the body.

At about the same time, another spate of wolf attacks occurred between February and August 1981 around the town of Hazaribagh, in the eastern India district of Bihar. This pack consisted of five wolves and their hunting range was a 2.7 sq miles (7 sq km) area around the town. They were attracted by the rich pickings at the town's rubbish dump, where livestock carcasses and human bodies unclaimed from the local mortuary were not buried properly and were dug up by wolves, hyenas, jackals and domestic dogs. On one night in June 1981, five wolves were picked out by a spotlight, and were seen to excavate and eat human bodies. Whether they acquired a taste for human flesh, we will never know, but local children began to be targeted.

One of the first attacks was on 15 February 1981, when a wolf entered a yard fenced by brushwood and attacked a young boy. Several people heard the commotion and came running. Taking wooden poles, they beat the wolf to death. During the next six months, wolves were to kill and eat 13 children aged between four and ten, and a further 13 had close calls but mercifully escaped injury or death. During the year four wolves were killed in the vicinity of Hazaribagh, and at the end of this period a large male, thought to have been the alpha male of the pack, was caught in ravined scrub forest and shot. The villagers thought this marked the end of the attacks, but about four months later they were to be proved wrong.

At dusk on 21 December 1981, a wolf grabbed a seven-year-old boy from a compound and ran into the scrub. His mother and a relative raised the alarm, and neighbours eventually headed off the wolf. Seeing that he was surrounded, the wolf dropped the boy and made his escape. Five days after the attack, the wolf researcher S. P. Shahi went to the site of the attack and also visited the hospital where the boy was being treated. At the attack site, the unmistakable pug marks of an animal with non-retractile claws, like those of a wolf, were found. Where the wolf had carried the boy, its large canine teeth had punctured the child's abdomen and left flank, the length of the bite marks consistent with an attack by a dog-like predator, such as a wolf, rather than a large cat, such as a leopard. Whether wolves were the actual perpetrators of the attacks is not clear. Other dog-like animals could have been responsible, such as feral dogs (*Canis lupus familiaris*) and dholes (*Cuon alpinus*).

The Indian wild dog or dhole (meaning recklessness or daring) is not a true dog at all. It is more closely related to our red fox, although considerably larger and more powerful. Like the wolf, it is accused of killing and eating people, but the evidence is slim.

The accounts of children as wolf victims in India, however, have not stopped. In April 1985, newspaper reports brought news that wolves had killed three more children in eastern India. And in June 1986, forestry officials in Bihar offered a Rs5,000 (£250) reward for the capture of wolves that had killed more children in the Hazirabagh region. Wolves, according to press reports, accounted for the deaths of a further six people by November.

In 1996, another spate of attacks occurred in a 60-mile (97km) stretch of the Ganges river basin in Uttar Pradesh.

According to police figures, between April and September, 33 children were taken by wolves and another 20 seriously mauled. At the village of Banbirpur, according to a report in the *New York Times*, a woman was taking her eight children to a grassy clearing which was used as a toilet when a wolf grabbed the youngest, a four-year-old boy. It carried him into the corn and elephant grass that lined the river bank. Only his head was found three days later. Claw and tooth marks indicated the assailant was a dog-like animal, probably a wolf.

Local rumours are rife. Some believe the attacks are by werewolves, others blame Pakistani infiltrators dressed as wolves. Banbirpur villagers, who live in one of the poorest and over-populated parts of India, even tell of the predator approaching on all fours like a wolf but rising up on two feet to make the attack. They describe a person in a black coat, helmet and goggles! Despite the bizarre stories, fairy tales and folklore, the local authorities are convinced the attacks are by wolves, and it would not have been the first time. In 1878, British officials reported 624 human deaths from wolf attacks in the same area. In that year, 2,600 wolves paid the ultimate price – they were shot, and the killings stopped.

Elsewhere in India, predation on humans by wolves is almost unknown. Local people living in and around the Melekote wolf sanctuary, Gujarat in the Rann of Cutch, and the Mahhuandanr Valley of Bihar – all wolf habitats – consider wolves to be an 'occupational hazard'. They take sheep but never children. Indeed, some communities welcome the wolves for they keep crop-raiders, such as wild boar, under control. The Indian grey wolf, however, is an endangered subspecies and despite these localised attacks on children and the taking of domestic livestock, the general feeling in India is that it should be protected.

Like the other large predators, the wolf has been subjected to serious habitat loss. It was once very common in the wilder semi-arid regions of India, but now these living places have either been turned into deserts by the overgrazing of domestic livestock or have been modified by irrigation and taken over by agriculture. As a consequence the wolf is confined to isolated pockets of wilderness, and its natural prey – blackbuck, chinkara, wild boar piglets, hares, ground-living birds and, when times are hard, scarab beetles – has disappeared. The few wolf survivors now occupy sites close to where sheep are reared.

Similarly, the grey wolf in Italy is under threat, partly from the activities of hunters who see the wolf as a rival for game, and partly from the reproductive urges of feral dogs, that is, domestic dogs that have gone wild and reverted to the lifestyle of the wolf. In the Abruzzi Mountains, for example, the pure genetic strain of the wolf is being diluted by inter-breeding with feral dogs. Here, packs of feral dogs – with dogs the size of Pyrenees mountain dogs running alongside house-hold poodles – hunt like wolves. In the 1980s an estimated 80,000 feral dogs throughout Italy were fuelled by a constantly replenishing pool of ten times as many strays and free-ranging pets not under supervision. And this compared to a wolf population of no more than 150 animals.

The dogs travel and hunt at night in packs of 20–30 indi-viduals, the larger breeds, such as mastiffs, German shepherds, and setters, dominating the groups. Rather than scavenge exclusively on human rubbish, however, they have been taking over the ecological niche left vacant by the decline in Europe's larger predators, such as the wolf, bear and lynx. They have become real hunters, yet with a competitive advantage – they hunt like wolves, but look like dogs – and so, when approaching rubbish dumps or flocks of sheep, they are less likely to be shot on sight.

Actual attacks on humans are rare: they generally avoid people. But there have been hints that feral dog packs could attack a solitary walker. Luigi Boitani once told me of a time when he was coming down from the mountains and was tracked by a pack of feral dogs, an occasion during which the good professor felt distinctly uncomfortable. The dogs scavenge mostly at rubbish dumps but, given the opportunity, will try to bring down sheep, cattle and even horses. Could people be on a future menu?

In 1956, a postman whose round was a couple of miles from the city of Rome was killed and partially eaten. At the time, people attributed the attack to a lone wolf, but might it have been a feral dog or even a wolf-dog hybrid? Elsewhere feral dogs are certainly a danger to humans. In the winter of 1985, a pack of feral dogs attacked and killed a man on the outskirts of Belgrade, in the former Yugoslavia. Twenty-three hunters scoured the suburbs and exterminated the entire pack of about 50 animals.

Bucharest in Romania has become the city of the dogs. An estimated 100,000 to 200,000 dogs roam the city, biting 50 people a day. The dogs were released in the 1980s, after Nicolae Ceausescu's building spree in the city. Displaced people were moved into modern apartments but they abandoned their dogs. The dogs bred and today there is a plague. So far, dog attacks have not been serious, and they do not view people as food.

In the USA, however, there are reports of feral dogs attacking people with the obvious intention of feeding on them. The New York borough of Queens recently had a pack roaming the streets. It was led by a pair of German shepherds and it attacked humans daily. People walked the streets armed to the teeth, but the pack was not eliminated until its den was found in an abandoned building and the dogs were killed. A pack in another US city, containing a beagle, a dachshund and several breeds of terrier, brought down and killed an 80-year-old woman. And during 1998, a small pack of stray dogs attacked and injured a Massachusetts boy on his way to school, and mailmen were attacked in Oklahoma City. Feral dogs also roam the countryside. In October 1995, a young boy was fishing at Buna, Texas, when he was attacked by a pack. He rolled himself into a ball, protected his head with his arms, and survived.

In fact, dogs appear to be far more dangerous to people than are wild wolves. Rottweilers, lurchers and pit bull terriers – all working dogs – have been guilty of attacking many people. There have been many badly mauled children and several deaths. In 1984, a pack of starving German shepherd dogs attacked two eight-year-old girls in the suburbs of Turin. The dogs had been left to guard business premises over the August holiday period but had no food or water. Driven by hunger, they squeezed through a gap in the perimeter fence and slipped out of their yard. Despite attempts by the father to drive back the dogs, they killed one of the girls. Eventually, sharpshooters shot the dogs.

In January 1984, an 80-year-old woman in Germany was attacked and partly eaten by her 18 greyhounds. Most probably she died naturally and was eaten by the starving dogs. Another German dog victim, a 26-year-old girl, who lived in Bonn, was attacked and killed by the Alsatian she was walking for her uncle.

In Britain, the pattern is repeated. In April 1999, a Japanese Akita hunting dog attacked a three-year-old girl in Sunderland, grabbing her by the face and throat. The girl suffered serious facial injuries and required skin grafts but survived the attack by the 140lb (64kg) dog. In March 1999, a family pit bull killed a 14-month-old toddler in the backyard of their home. The dog had ripped the flesh off the child's face. And in August 1999 a four-year-old boy needed 136 stitches in the head and face when a Rottweiler escaped from the garden of a house in Caerphilly, South Wales. The boy's face was streaming with blood and his face and head pitted with teeth marks. In each case the immortal words were heard – 'It was the first time the animal had attacked anyone'.

It is in the USA nevertheless, that the dog-attack problem has reached almost epidemic proportions. In New York City, statistics for 1987 showed that in a year when there were only 13 shark-bite incidents around American coasts, there were 8,064 cases of dog bites in the city. Some were serious. In fact, the number of serious attacks had increased to such an extent that in January 1999 insurance companies in the USA were considering dog attacks to be a home-owner's greatest liability and some were beginning to drop pit bull and Rottweiler owners from their books.

In Michigan – fifth in the USA in fatal dog attacks after California, Texas, Alaska and Florida – recorded 13 deaths from 1987 to 1997. Statistics prepared by the Michigan Association of Insurance Agents in 1997 were startling:
- In a 12-month period during 1995 and 1996 at least 25 people died from dog attacks throughout the USA – 20 of the fatalities were children.

- Between 1980 and 1994, dog attacks increased by 37 per cent while dog ownership increased by only 2 per cent.
- About 70 per cent of dog attacks occurred on the owner's property.

In another survey by the Center for Disease Control and Prevention in Atlanta, it was revealed that 4.7 million people or 2 per cent of the population of the USA are attacked by dogs each year. Since 1994, those requiring medical attention rose by 37 per cent. Pre-teens (10–12-year-old children) are most at risk, with nearly half of all attacks on this age group. A typical case was a boy waiting for the school bus who was attacked by three Rottweilers. The dogs tore him apart in front of his younger brother and 17 classmates. His spinal cord was crushed and the carotid artery and jugular vein severed. By the time the police arrived, the boy's body was empty of blood.

A third survey by University of Pittsburgh researchers in 1998 revealed that in the USA 334,000 victims of dog bites need hospital treatment and there are 20 deaths each year. The total annual medical bill is $102.4 million.

Autumn 1995 was a particularly bleak period. In September, a pet chow killed and partially ate a two-week-old baby girl; in November a four-year-old boy was taken out by two Rottweilers and a five-year-old girl was attacked by a Rhodesian ridgeback and a pit bull; and in December a six-year-old boy was attacked by a pair of German shepherds and a 90-year-old woman was brought down by two pit bulls. The following month there was more mayhem. A chow bit off most of a five-year-old boy's face, a Rottweiler dragged another five-year-old around by the head, a nine-year-old girl nearly lost an arm to a pit bull, a chow dragged a three-year old off his tricycle and chewed his head, and a Great Dane bit the face and punctured the skull of a ten-year old boy. And the carnage goes on.

In May 1998, two pit bull dogs terrorised a neighbourhood in Charleston, West Virginia. The dogs were near a school where the school secretary took refuge in her car and a man leaped on to the roof of his car to escape. The dogs killed a smaller dog before police shot them. Fortunately no children were harmed, unlike another incident in October 1998 at Atlantic City, New Jersey, when three pit bulls actually attacked a group of children playing in a schoolyard, biting them on the buttocks, legs, hands and backs. School teachers and others who were passing tried to get the dogs off and were bitten for their trouble. Two of the dogs were caught by police but the third escaped. Nobody was seriously hurt.

In September 1998, three large dogs – two mastiffs and a Saint Bernard – killed an eight-year-old boy in Durham, North Carolina. He had climbed over a garden fence in order to visit a neighbour, an eight-year-old girl. As the boy jumped down, the dogs attacked. The girl ran to his rescue, putting her body between him and the dogs. She was covered in blood, according to police. The boy's chest was punctured at least twice, and he had serious bites to the neck and arms. The girl was untouched, but the boy later died in the hospital.

Yet not all attacks are from large and obviously potentially dangerous dogs. No less an organ than *The Journal of Paediatrics* published the case of a child-killing Yorkshire terrier.

Nevertheless, pit bulls, Rottweilers, shepherds (Alsatians), huskies, malamutes and Dobermans lead the list of dogs involved in fatal attacks on children, with chows, Great Danes, Saint Bernards and Akitas coming a close second. But in the USA today, a disturbing trend in dog breeding is presenting us with another potentially dangerous scenario. It comes in the form of a deliberate cross between the wolf and a domestic dog. To the trained eye the dog differs from the wolf in having taller ears and a longer snout. There are also important behavioural differences. While decades of selective breeding has 'tamed' the domestic dog, a few generations of domestication has failed to breed out the wolf's basic predatory instincts. To a wolf-dog hybrid, a small child can looking enticingly edible; and anyone playing, screaming and running can resemble prey in distress, a primary target for any predator. In an instant, the wolf's normal instincts can trip in and a relatively innocuous wolf-like dog may turn to man-eating.

So, the keeping of wolf-dog 'hybrids' as pets has resulted in several gruesome attacks. On 2 March 1989, a five-year-old girl stepped off her school bus at Upper Peninsula, Detroit, and while walking the 100 yards (100m) to her home was attacked, killed and partly eaten by a dog that was part malamute and part wolf. Not far away in Rothbury, to the north of Muskegon, a 12-year-old girl was waiting at the school bus-stop when she was attacked by a cross between a wolf and a shepherd on 21 October 1996. The dog pressed home its attack for about 15 minutes and the girl was eventually rescued by the school bus driver, but not before the dog made four deep bite wounds to her head, damaged her left hand, and removed chunks of flesh from her right arm. It took seven hours of surgery to sew her up again.

In April 1991, a wolf-dog hybrid severed the arm of a 16-month old child in New Jersey. At Fort Walton Beach, Florida, a wolf-dog hybrid was advertised as 'pet of the week' and within hours of it being taken home it had killed a neighbour's child. This desire for the macho image had resulted in six children being killed by wolf-dog hybrids during the previous three years in the USA.

Some people keep real wolves and the outcome is inevitably the same. In April 1999 a pet wolf in Bangor, North Down, in Northern Ireland, attacked a small girl who was visiting the house. The child was on the floor playing while the wolf was sleeping. Suddenly, it leapt up and grabbed the child by the head. Apart from a few puncture wounds in the lip and scratches to the nose and forehead, the girl was safe.

Clearly, the more appropriate place for a wild dog, such as a wolf, is not in the home but in the wilderness, yet even seemingly harmless wild species have been implicated as part-time people killers. The coyote (Canis latrans) of North America, for example, is not whiter than white. Several cases are on record of coyotes attacking and eating babies and young children, and savaging adults.

In Yellowstone Park, in 1959 a coyote attacked a group of Boy Scouts who were out hiking near Slough Creek. One of the party had a penknife which was used effectively to see the creature off. A few years later, in 1974, a coyote attacked a television crew filming the geyser eruptions. And in 1990, a young man was skiing in the park when he was attacked. He was able to defend himself with his skis. In Canada in 1988, a coyote attacked an 18-month-old baby in British Columbia, and in a separate incident a coyote tried to pull a teenager out of a tent in the Banff National Park. A coyote in Griffith Park, Los Angeles, savaged three adults before it attacked and was clearly trying to eat a 15-month-old baby playing in the park in July 1995. The following month another coyote attacked a 16-month-old baby boy at Los Alamos, New Mexico, and tried to drag him into the brush. The mother ran screaming into the yard and frightened the animal away. It was the second coyote attack in the area within a month.

Attacks from coyotes have also been registered in Vermont, where in two quite separate incidents hunters were attacked in 1992; in Barnstable, Massachusetts, where a woman was bitten in 1994; and in 1995, when hikers were attacked at Mt Sunapee, New Hampshire.

In the southern half of Africa, the African hunting dog (*Lycaon pictus*) has been implicated in attacks on people in the bush, but these occasions are certainly rare. The dogs would have to be sufficiently hungry and desperate to mount such an attack, but stories exist of packs being bold enough to pull down solitary travellers.

The dingo or wild dog of Australia (*Canis dingo*) has also been implicated in attacks on people. In February 1999, for example, tourists on Frazer Island were reported to have been attacked by dingoes, the result probably of people offering titbits to the island's dingo population. The worst incident was an attack on a 13-month-old baby in April 1998, which provoked the government to consider a cull.

# CHAPTER 3

---

# MAN-EATING TIGERS

The tiger (*Panthera tigris*) is the largest of the big cats and a formidable predator. With a head and body length up to 10.7ft (2.8m), height at the shoulder up to 3.6ft (1.1m), and a weight of up to 660lb (300kg), the tiger is a powerful animal. Most of its strength is in the forequarters, with which it can push prey to the ground. Compared to most other big cats, the rear part of its body is less sturdy and so, instead of running prey down, it must manoeuvre itself as close as possible before launching an attack. It relies on stealth and concealment.

The tiger stalks its victim, approaching from behind or the side. It relies more on eyesight and hearing than its other senses, ensuring that it can see the target while the target cannot see it. Approaching in a crouched or semi-crouched position, it will try to get within about 66ft (20m), and then in an explosive burst of energy, it rushes the prey. Its principal weapons are 2in (10cm) long, retractile claws with which it can grab and hold its prey and four large canine teeth with which it pierces the neck or head of its victim. Canines in the upper jaw can be up to 3in (7.6cm) long and are partly encased in jawbone. In each cheek, the row of three carnassial teeth is used for tearing and cutting flesh, and the tiger's tongue is like a rasp quite capable of scraping off flesh. On contact with prey, the powerful forelimbs and sharp claws seize it by the shoulder, back or neck and it is forced to the ground. Unless the prey is very large, the tiger keeps its hind legs on the ground. Death can be either by a suffocating stranglehold to the victim's throat or a bite to the back of the neck, close to the base of the skull. This severs the vertebrae and compresses the spinal cord. Once caught and subdued, the prey is usually dragged into cover, sometimes clear of the ground.

Although tigers sometimes gorge themselves with a 44–77lb (20–35kg) meal, a carcass may not be eaten at one sitting. Tigers tend to eat about 33–40lb (15–18kg) per day, starting with the rump, and may return to a large carcass several times at three- to six-day intervals before it is finished. Left-overs are concealed with vegetation and soil to protect it from scavengers such as vultures and jackals.

A healthy female tiger without cubs may kill every eight days or so – about 45 kills per annum – increasing the frequency to every five or six days or 70 or more kills per year when accompanied by 6–10-month-old cubs. Tigers, like many other top predators, are not actually the most efficient killers. On average, they only make one successful attack for every 20 attempts. Alert, therefore, to any feeding opportunity, they have been seen to chase leopards from their kills, and will take livestock.

Tigers sometimes hunt co-operatively, with families acting together. Tiger conservationist Valmic Thapar describes how he watched two males and three females hunting like lions. The group – possibly a family group – took up positions around a lake and drove prey from one to another until one tiger was successful in bringing it down. Attacks on humans, however, tend to be by solitary animals. Lone tigers have been eating humans since people and tigers have shared the same forests. Records of tiger attacks on people go back many hundreds of years. A mural in stone at Humpi, dated 14th century AD, depicts an Indian man being attacked by a tiger. The animal is shown rising up on its hind legs.

Statisticians have estimated that during the last 400 years over a million people throughout southern Asia have been killed by tigers – an average of over 2,500 a year. In 1769, records show that 400 people were killed by tigers in an area around Bhiwapur in India. In 1821, one district in the foothills of the Himalayas reported 373 deaths from tiger attacks. Man-eating tigers were so numerous, according to one British colonialist and hunter, it was in question whether man or the tiger would survive. Many tombstones of the British dead in India have the epitaph, 'Died of injuries received from a tiger'. At the turn of the century the British authorities in India kept records of fatalities due to wild animals. From 1902 to 1910, the statistics relating to deaths from tiger attacks were understandably disturbing – 1,046, 866, 786, 698, 793, 909, 896 and 882 – an average of 860 per year. During 1922, there were 1,603 deaths caused by tiger attacks in that year alone. In the Khandesh district, for example, there were reports of about 500 people and 20,000 cattle being killed and eaten by tigers. To put this into perspective, deaths due to snake bites averaged 21,000 a year.

Nevertheless, the tiger was always headline news. Sometimes it was quite bold. Colonel Keshari Singh – the so-called 'tiger of Rajasthan' – recalls the story of a newly married girl who was sleeping beside her husband in a locked room when a tiger somehow broke in and took her away.

The tiger, however, did not always get its man. Robert Sterndale recalls an incident when tracking a man-eater in India. He came to a hillside on which cattle and buffalo were grazing, when the cry 'Bagh!, Bagh!' went up. The cattle stampeded, and the sound of a struggle between a herdsman and the tiger could be heard coming from a stand of trees. The buffalo did not join the cattle. Instead they threw back their heads and, in a tightly bunched group, rushed the spot where the tiger attack was under way. As they charged through the brushwood, herdsmen and village folk came running up and carried the injured man back to the village. He was badly scratched but was not seriously hurt. Having seen the tiger racing towards him, he had stuffed his blanket into its mouth and then hit it continuously with an axe. Such was the strength of the tiger, however, that if the buffaloes had not charged, the man would have been killed. But they saved the day.

One of the most striking man-eaters was a black (melanistic) tiger that frequented the countryside around Chittagong in the 1840s. It killed cattle at first, but then turned to man-eating. It attacked and partly ate a villager about 3 miles (5km) to the south of a military base there. It was found eventually by the roadside by a Fellow of the Royal Zoological Society, one C. T. Buckland. It had been hit by a poisoned arrow.

It was Buckland who noticed, all those years ago in the mid-1800s, that man-eaters were not necessarily the old, deformed or mangy animals that were usually considered 'man-eaters'. He saw a tigress, estimated to be about five years old, that had started to prey on villagers near Dacca. Her skin was 'clean and well-marked', and she looked in rude health.

Considering the number of people in the region and their close proximity to the population of tigers living there, it has surprised scientists that tigers have not killed more. But when they do kill, entire communities live in fear.

The man-eating record is held by a tigress which killed 438 people over a period of eight years in the early 1900s. Its first hunting ground was in Nepal where 204 people were killed, and then it went over the border to Kumaon where it claimed a further 234 victims. It gained notoriety as the 'Champawat Man-eater'. Big-game hunter Jim Corbett was summoned to hunt the killer, and when he arrived at one village he found that all the people were so terrified that they had locked themselves in their houses for five days, not daring to come out. Outside, the tigress stood defiantly in the road, waiting for her next victim. She was killed by Corbett in 1911.

At about the same time, another tigress was injured when attacking a porcupine. She lost an eye and had 50 or more quills embedded in her right foot and foreleg. She could hardly move for the pain and lay starving, but she still managed to kill two people who chanced upon her. She failed to eat the first, but partly devoured the second, and then took to man-eating full-time. Twenty-five people lost their lives before Corbett was able to track her down and shoot her.

From encounters like these, Corbett was able to identify a likely reason for tigers turning into man-eaters. He wrote: 'A man-eating tiger is a tiger that has been compelled, through

33

stress of circumstances beyond its control, to adopt a diet alien to it.'

He also recognised that a man-eater is not necessarily an old tiger: 'A tiger, when killing its natural prey, depends for success of its attack on its speed and, to a lesser extent, on the condition of its teeth and claws. When therefore a tiger is suffering from one or more painful wounds, or when its teeth are missing or defective and its claws worn down and it is unable to catch the animals it has been accustomed to eating, it is drawn by necessity to killing human beings. The change-over from animal to human flesh is, I believe, in most cases accidental.'

A wound alone, though, is not sufficient to turn a normal tiger into a delinquent. Tiger conservationist Arjan Singh tells of a trapped tiger in Kheri that was jabbed repeatedly in the rump by a spear. It eventually escaped and died some five weeks later of septicaemia and starvation. It lived its final days in a well-populated area yet not once did it turn to man-eating. Similarly, a sub-adult tiger with a lung punctured by a porcu-pine quill died from starvation in a highly populated area without having attacked a single human being. Another tiger with a broken foreleg became a successful scavenger rather than a hunter and never killed a person. Singh's conclusion is that a tiger must suppress a strong urge to avoid contact with people before it will attack.

If all tigers turned to man-eating when they reached old-age, the number of humans consumed each year would be far greater than it is. And if incapacitation is a prerequisite to man-eating, why don't all injured or old tigers take to it? The problem is clearly related to local conditions. Jim Corbett's view of the reason tigers attack humans was appropriate to a time when the forests were pristine, game was plentiful, people were few and most tigers to become perpetual man-eaters were old or injured and unable to hunt effectively. This situation persisted for many thousands of years, but all that has changed, and rapidly. Forests were cut, the buffer zones of grass and scrub between agricultural land and wilderness disappeared, and tigers have been the victims of the resulting ecological crisis. So, while old man-eaters still appear from time to time, in more recent years man-eating tigers have picked up the habit for quite different reasons. Today, they appear to be most common in areas where forest habitats have been lost and natural prey is scarce. Link this with agricultural expansion and increasing confrontations between people and tigers, and you have the recipe for disaster.

Naturalist Kailash Sankhala explored tiger habitats in Bastar and confirmed that loss of habitat meant loss of natural foods. The local people were also not stockmen, so tigers had no choice: the only prey available was people. Forced by hunger, they will pick off the last man in a file of workers walking home or attack a group of women bathing in the river. So terrified are the villagers that they abandon their homes. This happened in Abujmar when the only person to remain behind was the local schoolteacher.

At a village in the forests near Hyderabad, the only way the inhabitants discovered that their postman had been killed by a tiger was when a pile of unopened letters was found beside a pair of human skulls.

Interestingly, Sankhala noticed that people bending down to cut grass or gather crops were more vulnerable to tiger attacks than people standing up. During his time as director of Delhi Zoo, he tried some simple tests and found that when somebody was bent over, a tiger would begin to stalk. As soon as the person stood up, the animal would lose interest. Thus, tiger attacks on people are not thought to be the animal's natural behaviour. In some cases it could just be a case of mistaken identity. It seems that people who are standing up are not thought to be obvious prey for a tiger; they are probably seen more as a threat or a rival, than as prey. But someone bending down, say, to collect wood for the cooking fire, cutting grass for thatch or sitting on the ground, has the uncanny resemblance to a tiger's more usual four-footed prey and therefore presents it with a legitimate target.

Tigers and people have co-existed in India for centuries (carvings from the Indus Valley showing captive tigers date to 2500BC) without incident, but just occasionally an individual will turn man-eater. Peter Jackson, big-cat champion and tiger supporter, recalls meeting a tiger on a forest trail without being alarmed. The animal hid in bushes and waited for Jackson to pass by before continuing on its way. More often than not, this is the way the majority of villagers relate to tigers. Tens of thousands of people go about their daily business in their company. Villagers have lived alongside tigers, meeting them in everyday life without fear, but then suddenly one tiger will turn nasty. Rhino and bears are more unpredictable, and are more prone to attacks, yet when a tiger turns people-killer, the community is up in arms.

Revenge attacks, despite the tiger being a protected species, are common. In 1944, an Indian Forest Service manager killed a supposed man-eater in the Haldwani forests. This followed an attack on a man cutting grass in the forest. Only one of the man's legs was recovered, and the killing was clearly the work of a tiger, but which one? Hunters killed six more tigers living in the area, each man claiming the man-eater to be his. The reward was a free permit to hunt for a further year, but had any of them bagged the man-eater? Nobody knows, and at least five innocent tigers had been killed in an attempt to eliminate the guilty one.

During 1988, three people were killed near the Dudhwa National Park. As a consequence, 23 tigers were poisoned, electrocuted, trapped, shot or blown up by bombs implanted in deer carcasses. The park had been under increasing pressure from the surrounding human populations. Natural habitats had been replaced by sugar-cane plantations, not in itself a bad thing as tall sugar-cane is a suitable tiger habitat-substitute for long grass. But the wildlife has gone and with it the tiger's natural prey. Metal roads in the area enabled the illicit felling and taking of trees and the illegal killing of game. Poaching has accounted, for example, for many deer – the tiger's natural prey. Swamp deer populations (for which the park had become well known since Billy Arjan Singh's conservation pro-gramme) dropped sharply. The local population of tigers had no choice but to seek food elsewhere. Farm livestock outside the park was an obvious target, and their human keepers were an easy catch too. With people an increasingly familiar part of the landscape, tigers have gradually accepted them as 'natural' and therefore fair game, i.e. normal prey.

Whatever the reason, the pattern of attacks is becoming a familiar one. First the tiger kills livestock, such as cattle, and then as herdsmen try to chase it away, it turns and kills them. Realising how easy it is to kill people and having tasted the pork-like quality of human flesh, it repeats the exercise and the man-eater is born.

Along the border between India and south-west Nepal, man-eating tigers are active, despite the drastic reduction in tiger numbers, and frequently reports appear in the newspapers and scientific journals of tigers attacking people. One exception was a tiger in western Nepal that ignored cattle, calves, goats, buffalo and even adult humans and attacked only children. All its victims – 35 children from seven villages in the district of Baiditi about 250 miles (402km) west of Kathmandu – were aged under ten years old. The animal pounced on youngsters who were tending their parents' cattle or who came out of their homes in the evening. Army marksmen eventually shot it.

Many tigers moved out of Nepal when the forests were cut down and into the forests of northern India. When all the home ranges were occupied, they spilled over into the sugar-cane plantations on the newly cleared land bordering the forests. The sugar-cane fields were ideal places for tigers to hide. They stood all year round and were thick and secure. But they were tended by farmers, and so people and tigers came increasingly into contact.

In February 1967, for example, a group of 14 forest workers were out marking trees in the Corbett National Park in Uttar Pradesh, on India's border with Nepal, when one of them was seized by a tigress and carried into the bush. Shikaris (professional hunters) found the body and, moving it to open ground, set up a *machan* or shooting platform and waited for the culprit to return to its kill. She was shot that evening. Examination of the body revealed that two of her canine teeth had been broken off, an upper molar tooth was loose and four incisors were missing. She was an old animal, but the unfortunate man was her first victim.

In 1976, an outbreak of man-eating occurred in the Kheri District of Uttar Pradesh. About 138 people are known to have died during the following six years. Ten tigers were killed as a consequence, one still with human remains in its stomach; two were captured and taken to zoos where their health declined and they quickly succumbed to diseases.

Kheri was once a swamp area where mosquitoes bred and malaria was common. It was largely uninhabited. But with the coming of DDT, and the control of the malaria-carrying mosquitoes, the area has been settled by farmers and ranchers. The forests were pulled down, leaving small remnants, such as the Dudhwa National Park, for tigers to live. Frequently, they would leave the park in search of food, and 200 people were killed during a ten-year period from 1978 to 1988.

In August 1989, another spate of attacks drew attention to the Kheri region. A tiger killed three people on the outskirts of Kirtinagar. Loss of natural habitat was blamed. Since then, half the human kills have occurred in the sugar-cane fields, the very places favoured by tigers – a substitute for their normal tall grass habitat. It is likely, therefore, that people will come into contact with a resting tiger, which may attack partly in self defence (female tigers use the fields as birthing sites and are thus naturally aggressive towards intruders) and partly because it has been presented with an easy target.

The problems with sugar-cane became self-evident in a series of incidents in 1983 that took place at Ghola, an area of buffer land outside the park near Sathiana. The area had been cleared and sugar-cane planted right up to the park boundary. One May night a tiger pounced on a night-watchman guarding the crop, but a lantern overturned during the struggle, setting light to some straw, and the attacker made off. The tiger returned during the day, however, and seized the man in broad daylight while he was stripping cane. The tiger bit into his head, but the man had some cloth wrapped around it and the wounds, though serious, were not fatal. He had survived two attacks. The third attack did not end so fortunately. A man visiting friends in the area squatted down beside the sugar-cane to relieve himself when a tiger grabbed him and dragged him into the field. His friends set light to the crop in an attempt to frighten the tiger away, but when they found the man's body he was already dead. The local population was up in arms, demanding that all the tigers in the area be killed. The following morning, there was a fourth attack. A girl was taken from just outside her hut and hauled away. By the time relatives had tracked her down, she was also dead. A few days later a man was tending his field next to a patch of long grass and a stand of sugar-cane when a tiger leapt out and tried to drag him away. The man yelled and his neighbours came running. The tiger dropped him but it was too late, for he died sometime later.

Again the community was up in arms, demanding something be done. The director of the Dudhwa National Park arrived to hunt down the killer. He put up a shooting platform in a nearby pipal tree and tied a live bait below. But at the moment the tiger came and killed the bait, the director was elsewhere. The next night the tiger returned to finish its meal but the remains had been tied up carelessly and so it was able to run off with the carcass. The director fired but missed. The tiger ran off, returning yet again to retrieve its food later that night. By this time the local people felt particularly vulnerable. The tiger had killed five people, but had eaten none of them. It would surely attack again. Another bait was tied below a *machan*. The tiger charged, the director fired ... but only succeeded in winging it. So, when Arjan Singh (an ex-tiger hunter but now a dedicated conservationist) was asked to take over the hunt, he had the doubtful pleasure of trying to track and subdue a wounded tiger.

Four elephants were used to comb the sugar-cane field. Visibility was zero, but splashes of blood and pug-marks indicated that the tiger had been shot in the foot. Suddenly, the tiger snarled: it was close to one of the elephants. The hunters could not decide what to do, and in the ensuing argument, Singh implored the others to press on with finishing off the tiger. Unable to give orders, he stepped down as temporary leader. Things then became even more ludicrous. One of the hunters suggested getting a tractor to pull a plank through the cane to create avenues across which the tiger would inevitably move, but the machine could not arrive for a couple of hours. In the meantime, the owner of the field arrived and took umbrage at the way his field was being destroyed. He argued with one of the hunters, the hunter let off a barrel of his shot gun, and the owner drove off in a huff. The local villagers burned his sugar-cane as a reprisal. As the flames rose, it was clear that the tiger would eventually emerge from the field where the blaze was likely to burn out. The director, having returned to the fold, lined up his marksmen. Three .375 magnum rifles, one .30-06, two '315s and some smoothbores waited to greet the limping tiger, together with about 200 local inhabitants. This time they didn't miss.

After events such as this, steps were taken to discourage sugar-cane production close to national parks, and buffer zones were created to separate tigers from people. As a consequence, there was an unexpected lull in 1984, when record low figures were announced. In that year just nine people lost their lives to tigers.

Although many tigers have moved into northern India, some have remained in the only surviving refuge for tigers in Nepal – the Royal Chitwan National Park. Those that stayed within the park boundaries were protected and followed a relatively normal existence, but the problem was the people living around the park. Without a buffer zone between tiger forests and agricultural land, and the incursion of people from settlements close to the park boundaries, tigers with home ranges close to the edge of the park have had more contact with people and livestock, and the inevitable has happened.

Between October 1980 and May 1988, 16 people were killed and partly eaten by tigers in and around the park. In 1984, there were three deaths and many instances of people being mauled. In June 1984, two villagers were killed near the park airport at Meghauli, and a week later a tigress ambushed an employee at one of the tourist lodges in the park. Tigers were concentrating along the fringes of the park where they were finding cattle an easy target. Occasionally people were killed and eaten too. The baiting of tigers for tourists to view them at Tiger Tops Jungle Lodge in Chitwan was considered by some people to be linked to attacks on humans, but tiger biologists, like Chuck MacDougal, David Smith and Hemanta Mishra, were not convinced.

More probably, the Chitwan tigers attack people when their

more usual food is unavailable or too difficult to catch, as when an animal is injured. Such was the case of a tiger at Chitwan in 1978 . After it left its mother's area, at about two years old, it was badly injured on the upper left leg in a fight with a larger male tiger over the possession of a slice of territory. Park rangers found the injured animal, darted it, cleaned up the wounds and fitted a radio-collar. But the lame tiger was at a significant disadvantage compared to other healthy individuals, and it settled at the edge of the park where it relied on cattle and domestic buffalo for food.

One morning in December 1978, however, the schoolteacher from the Nepalese village of Trilochan was going to the river to bathe. At the same time, the tiger was coming up a gully close to the river in order to take refuge in nearby woods. As the man reached the top of a small rise by the river bank, the two met, and the tiger did what was natural – it killed an easy prey. Another man, walking about 50 yards (50m) behind the schoolteacher heard a scream and then saw the attack.

Next day, Dave Smith, from the University of Minnesota, one of the Smithsonian Tiger Ecology group, who had previously darted and radio-collared both the cub which had turned man-eater and its mother, arrived from his camp nearby. Journalist Peter Jackson (chairman of the Cat Specialist Group of the International Union for the Conservation of Nature and Natural Resources) was with him, and he recalls how the villagers were understandably angry and frightened. They shouted at the research team that they would be put in prison if they killed a tiger, but if the tiger killed one of them, they were just dogs. The sentiment echoed those from elsewhere on the Indian subcontinent where a growing tiger population is coming into conflict with a rising human population. 'Are tigers more important than humans?', the people asked.

In an area where attacks on humans by elephants, bears and poisonous snakes are common, these responses to tiger attacks were disturbing. After all, in order to conserve tigers, the researchers need the co-operation of the local people. As a consequence man-eaters are dealt with quickly. On this occasion, the research team was able to track the radio-

41

collared tiger, dart him and take him to Kathmandu Zoo, the repository for delinquent tigers.

A similar situation involved the tiger Bange Bhale, otherwise known as 'bent foot'. In essence, his story describes the making of a man-eater. It is not known how he received his deformed right forefoot but unsubstantiated evidence suggests he was involved in a fight with a neighbour. Bange Bhale appeared at Chitwan in mid-1981. His arrival coincided with the disappearance of the resident male in an area to the west of Tiger Tops, between Dhakre and Ledaghat. Bange Bhale took over the territory, and with it the two females whose home ranges overlapped with his. Twice in 1982, he became a father. In 1983 he sired another cub, a female which was given the name Three Toes. Unlike her mother, Three Toes did not succumb to the allure of Bange Bhale, but took a fancy to his neighbour, a male the park officials called Lucki Bhale. In the meantime, another interloper, Bahadur Bhale, moved into Bange Bhale's territory. In March 1984, David Smith was in a canoe on the Rapti River, when he saw two tigers fighting furiously. One leapt into the river and limped away: it was Bange Bhale. After the confrontation, he immediately lost half his territory.

For the older tiger, it was the turning point. On two occasions, park rangers saw him being chased from kills on the eastern side of his territory by another neighbour, Kanccha Bhale. Disaster befell disaster and eventually Bange Bhale lost his entire territory to his western neighbours, and all the young cubs he sired were killed. By May 1984, he had left the immediate area and was wandering alone in the forest at the edge of the park. On the morning of 25 May 1984, wracked with hunger and handicapped with a damaged leg, he killed his first human victim. A man was cutting grass inside the boundaries of the national park, not far from Tiger Tops, and the injured tiger pounced on him and partly ate him.

Bange Bhale must have travelled some distance without eating, but on 2 June the tiger clocked up its second attack on a person at Nawalpur. A third was to join the list of human fatalities in August. The tiger was not in the headlines again for several months, but on 29 December he came across a mahout and his assistant gathering leaves for their elephant. The assistant shinned up the tree, while the mahout gathered fallen leaves on the ground. Bange Bhale spotted the assistant and began to stalk him, but then discovered a far easier target directly ahead on the ground. He grabbed the man, but the mahout had a bunch of leaves in one hand and a sharp sickle in the other. He thrust the leaves into the tiger's face and slashed its right cheek. The tiger fell, but immediately sprang up again, grabbing the mahout on the right side of the jaw, ripping his face badly. A tooth pierced the man's neck, but fortunately it missed his spinal cord. The man fought on and Bange Bhale was forced to retreat when two men approached with the elephant. The victim was raced to hospital at Patan and, apart from some paralysis on one side of the face, he recovered.

The hunt for Bange Bhale was on. His year, though, did not start well for he was shot with a tranquilliser dart by Hemanta Mishra, director of Nepal's King Mahendra Trust for Nature Conservation, on 1 January 1985 and was moved to Khatmandu Zoo. Curiously, the tiger was in good condition with no signs of his former injuries. He lived there until old age took its toll. On 31 January, however, baits were put out for another man-eater operating to the west of Tiger Tops. Could it have been one of Bange Bhale's offspring, or had Lucky been ousted from his territory just as he had expelled Bange Bhale and was forced to hunt for easier prey?

An old tiger was responsible for three fatal attacks on villagers in the Modhi Khola district, not far from Chitwan in the early months of 1986. It was tracked for a fortnight by two shakiris, working with the Smithsonian-Nepal Terai Ecology Project. The animal was shot, and an autopsy revealed it to be in very poor condition.

As has already been mentioned, not all attackers are old, sick or injured animals. Healthy tigers, like any predator, will take advantage of an easy meal. Tigers are hefty eaters. It has been estimated that a 9½ft (3m) long adult, weighing about 450lb (204kg), needs about 5,000lb (2,260kg) of meat per year. So, an individual must kill a large animal about once a week, all the more difficult when one considers that tigers fail to catch their prey in 19 out of 20 attempts.

A Bengal tiger's more usual prey may be sambar deer, hog deer, chital or spotted deer, muntjac, wild pigs and the occasional monkey – either grey langurs or brown macaques. From time to time, tigers will bring down even larger prey, such as gaur – the long-horned wild cattle of southern Asia – or even a baby one-horned rhinoceros. With the tiger's natural forest home rapidly disappearing through logging – legal and illegal – its prey is declining too. Domestic livestock and people are a handy substitute.

There is also the problem of overcrowding. With many males vying for the same plot, many more fights break out and tigers are injured. If the number of young tigers looking for home ranges in a reserve exceeds the number of ranges vacant, the crowded-out tigers are forced to live on the edges of the reserve. Inevitably, they chance upon livestock and its owners. This combination of more injured animals and greater contact with people can result in an increase in the number of tigers that turn to man-eating.

It is often the people, however, who are putting themselves at risk rather than a tiger behaving abnormally. Most deaths at Chitwan, for example, occur during the 15 days each year when local people are allowed to enter the park to cut grass. During 1987, for instance, there were several attacks. A group of villages were cutting roof thatch when they inadvertently surrounded a tiger hiding in the grass. As it tried to escape, it bit a 15-year-old girl on the head. One of the animal's canine teeth pierced her skull, yet she survived. Sometime later a women disturbed a tigress with cubs and the startled animal took a chunk out of her thigh. On the eastern side of the park the half-eaten body of a night watchman was discovered, and a women – also partly eaten – was found. Other people were mauled but escaped.

The following year the seasonal spate of killings continued. On 26 April 1988 a women grass-cutter was taken by a tiger, and again on the 16 May a man was killed. The victim's companions were no more than 50ft (15m) away in the tall grass, yet they failed to see the attacker. All they heard were the man's screams.

At Chitwan, the local tigers are not too fearful of humans.

Since the establishment of the park, its protection, and the popularity of Tree Tops, with its baiting sites and many tourists, tigers have become used to having people around. They will often stand and stare at a person, rather than run and hide, but rarely attack. It is quite possible that a grass-cutter bending over does not look like a person, and so is considered by the cat to be fair game.

Very rarely European visitors are the victims. This happened on 22 February 1985 when a British ornithologist was killed by a tiger in Corbett National Park, Northern India. He was leading a party of bird-watchers when he saw a forest eagle owl. He raced over a ridge into the thick jungle, straight into a waiting tiger. When he was discovered, his face and neck had been badly injured and a part of his left leg had already been eaten. Park wardens used elephants to recover his body.

At first, officials thought the culprit was a male tiger, named Dhittoo, which spent some time in that area. A buffalo bait was put out and a tigress – Diana – took it, dragging the body to the same spot where they found the man. She was considered to be the killer, but it was felt the killing was not her fault. She had small cubs, and simply took advantage of the unexpected prey in order to build up her fat reserves and so provide her offspring with sufficient milk. The park authorities kept an eye on the pair. In March, a woodcutter was attacked by a tiger but survived, and then a labourer was killed. A mahout taking a tourist party around the park found shoes, splashes of blood, and drag-marks on the ground. The next day, the half-eaten body was discovered. Nearby was Dhittoo. He was trapped, caged and transported to Lucknow Zoo.

A steady trickle of tiger-attack reports reach the newspapers each year. In November 1997, a tiger was reported to have killed more than 100 people, many of them children, during a 10-month period in the Baitadi district of western Nepal. It was tracked down and shot. The tiger, according to the terrified villagers, had taken to hiding under the roofs of houses. It would pounce on anybody who stepped outside.

One of the more recent incidents was at Sauraha, near

Chitwan. On 1 August 1998, a tiger took a man from inside his home. It crept silently into the hut, grabbed the man and his screams woke up his wife. She threw an oil lamp at the beast, but it just stood and stared at her momentarily before dragging the husband away. Two days later it returned to the same village. It tried to break into a house, but failed and killed a water buffalo calf in the adjacent stable. It then stalked a cyclist in broad daylight, but was scared away by a mahout on his elephant. It kept coming back, but took mainly live-stock. Tourists in a lodge nearby took to sleeping with their window shutters tightly closed even though the air tem-perature was soaring. Government officials arrived and the tiger was tranquillised and moved elsewhere.

Not all tigers who attack people are labelled 'man-eaters'. Some take to eating people for a short spell and then return to their normal diet. Such was the case with Chitwan's tiger 127. It was known to have killed four people in 1980–1, but then stopped. It reverted to naturally abundant prey, such as deer, and because it never killed another human being again, it lived the rest of its days in the wild.

Back in India, another man-eating saga occurred near Govindgarh in the Kaimore Hills of Madhya Pradesh, in the area once frequented by the white tigers of Rewa. In fact, a white tiger had been killed in the area in 1944, but since then, no tigers had been seen in the district until 1981 when a pair moved into the area and produced two litters of cubs – a male and female cub on each occasion. It was one of these cubs that was to turn man-eater. The more usual food was chital, wild boar and sambar, but game was becoming scarce. On 29 December 1986, people from the village of Papra, on the Shahdol to Umariya road, became the first victims of the infamous 'Man-eater of Papra'.

The first hint that something was not right came when a group of farmers were driving a herd of cattle into the Bhogan forest, some 2 miles (3km) from Papra. Suddenly, a tiger jumped out and attacked one of the cows. The herdsmen waved their arms and shouted loudly, and succeeded in chasing the tiger away. The cow died. The tiger returned and killed two more cows. Not long after, three women from the neighbouring village of Tara went to the forest to collect firewood. They split up and went separate ways. One of the woman had a dog and as they went deeper into the forest the dog barked at a bush. The tiger sprang out and grabbed the woman. Her companions came running, her screams and the urgent barking of the dog echoing through the forest. The two women were almost frozen with fear and could do nothing but wave their arms and scream at the tiger. Unimpressed, the tiger dragged the woman away, and all that was found later were a few leg bones.

For 12 months, the tiger terrorised the neighbourhood. Five more people were killed – three women and two men, and all within a radius of 16 miles (25km). On one occasion, a smaller party of men were in the forest when the tiger seized one. The others ran for help, but when they returned the victim was already dead and partly eaten. At this point, the authorities stepped in to catch the tiger. The Field Director of the Kanha Tiger Reserve, A. S. Parihar, arrived and set to work. His first task was to identify which of the four tigers known to be in the area was the man-killer. Was it injured, or incapacitated in a way which prevented it from catching normal prey? Baits of buffalo were placed at key sites in the forest, and watched. Tragedy struck again. On the morning of 24 February a women collecting firewood in the forest near Matwara was attacked and killed. When her body was recovered, the tell-tale canine marks were seen on her head and neck. A day or so later, two men were returning from a visit to the doctor and had stopped to collect firewood on the same hilltop where the woman had been killed. One heard the other scream, and ran to the spot. He saw his companion face down on the ground with a tiger standing over him. The victim was already dead. Pug-marks indicated the tiger was a large male.

For the next 25 days, Parihar tried to track down the man-eater. It killed cattle as well as people, but tracking it was difficult. There were few roads, the ground was rocky and hilly, and a hydro-electric dam building site meant that the area was disturbed by lorries and thousands of construction workers who were camped out in the hills. It was becoming clear why the tiger had turned to killing livestock and people.

Its natural prey had been frightened away. The next fatality was on 16 March. Accompanied by her two grandsons, aged nine and six, an old lady from Jhanjhar in the Bhander hills had walked to her field to tend her crops. She sent the two boys home to eat, and asked them to bring something back for her. When they returned, the woman was nowhere to be seen, so as night fell the boys went into the small hut at the field and waited for her to join them. One fell asleep, while the other remained awake. Suddenly he heard the snuffling of an animal outside the door. Thinking it to be a buffalo calf that had come to graze on their crops, he banged the door with a stick. The animal bolted.

In the morning, their grandmother had still not returned so they went to the village and raised the alarm. A group of about 50 villagers came to the field. They found blood marks in the doorway to the hut, and drag-marks on the ground. They followed the trail to the sandy bed of a *nullah* where a torn remnant of the woman's green sari was found. Police came and eventually found the woman's body. As they were about to move it, the tiger reappeared. Despite the presence of many people and the commotion they were causing, the tiger headed straight for its kill: it was clearly unafraid of humans. They managed to scare it away, and recovered the body. Pug-marks indicated it was the same tiger.

For the next few days, it went on another killing spree and six goats were taken. Two herdsmen from Jhanjhar were grazing their goats on a hillside when the tiger raced down the slope and killed the six in minutes. Understandably, the local farmers were afraid to tend their herds and fields. Some harvested their crops prematurely, before it was ripe. Footpaths and cart tracks fell into disrepair. People were afraid to travel. Parihar and his team were under great pressure to catch the beast.

The tiger meanwhile had moved to Gheri-ki-khoh. The site was staked out with baits – one in the *khoh*, one outside the village and another on a hill where the tiger had killed livestock. On 30 March, news came that the bait near the village had been taken. Parihar raced to the scene, and found that it had been dragged to a deep ravine nearby. Pug-marks indicated that the man-eater had been there. Then, scanning the rough terrain with binoculars, Parihar spotted the tiger hiding in the ravine. Placing a buffalo bait on the most likely route the tiger might take on leaving the ravine, Parihar and two companions waited on a platform for the tiger to appear. Mahouts on elephants approached the area from the opposite side of the ravine. If they saw the cat, they would signal. At 5.30 in the morning, three whistles indicated that the tiger had left its hiding place. More whistles told that it was heading towards Parihar's rifle party. They raised their weapons to their shoulders. The atmosphere was tense. They waited . . . and waited. A quarter of an hour passed, and then the shrill alarm calls of a troop of langurs in the distance announced that the tiger had moved on. It had escaped the trap.

The following night, they tried again. The bait – a live buffalo calf – stirred, and they raised their guns again. Each sound of the forest triggered the same reaction, but they were all false alarms. Then at 1.00 in the morning, a sambar bellowed, chital and barking deer called and a tiger snorted through its nose. The men were on instant alert. Peafowl and red jungle fowl called, breaking the tension, and the first rays of the sun accompanied the sounds of the village as it began to wake. The hunters had failed again. But just as they were ready to amble back to their billet, they heard the alarm call of a langur. It was very close. A leopard appeared momentarily, and then disappeared. Its presence explained the tiger's snort. Tigers make the sound to warn intruders to back off. Then, Parihar saw something move on the footpath. It was the man-eater. In one deft movement, Parihar grabbed his rifle, loaded it and placed it against his shoulder. His mouth was dry and the sound of his heart pounded in his ears. He aimed for the tiger's shoulder, squeezed the trigger, and the animal fell. But just as Parihar was lowering the gun, the tiger leapt up. His companion fired a covering shot but it missed, and the tiger ran into the cover of bushes. Now, they had an injured tiger on the loose.

Just then the elephants arrived and the hunting party took off after the wounded tiger. They spotted it at a water-hole, and tried to surround it. Another shot threw the animal to the ground, but it quickly recovered and slunk off into the forest, hiding below a stand of lantana bushes. Parihar fired at the spot, but his gun misfired. He tried again, and a bullet hit the tiger. One of the other hunters fired, and the animal was dead. The shooting brought the villagers to the killing site. Over 60 people came to help carry the dead tiger to Rajaha. Examination revealed that it had been a perfectly healthy male tiger in its prime. It was not injured, ill or old, but had taken to man-eating and stock-raiding because its natural prey had disappeared. It had been the behaviour of people rather aberrant behaviour on the part of the tiger that had caused the misery and terror. Habitat destruction had clearly caused the tiger to become a man-eater.

Elsewhere in India tiger attacks are less common, but occasionally still occur. In 1959, for example, an old woman was killed and partly eaten by a tiger when she was collecting myrobalan fruits in the forests in Karnataka, in the south-west part of the country. In April 1970, a woodcutter was mauled by a tigress with cubs. He fought her with an axe, but missed and fell over. The startled tigress made off, the man survived, but she was labelled a 'man-eater' nonetheless.

And there are often inconsistencies in controlling so-called 'man-eaters'. Kailash Sankhala recalls how, in 1971 near Rathambore, a tiger cub which had become separated from its mother was despatched by a volley of shots from police marksmen. It had an injured foot and was considered a potential danger to the public.

In December 1975 a young tiger that came out from the Sariska Sanctuary and approached a village in the Sikar district was the first to have been seen in those parts for about 40 years. The inhabitants panicked and climbed on to the roofs of their houses. The cub escaped from the pandemonium by hiding in a shed: but not for long – police sprayed it with a fusillade of light machine-gun fire.

Then there is corruption. Despite the absence of man-eating tigers in Rajasthan, the various postings of one particular official were always preceded by reports of man-eating. Shooting permits were issued to the official himself and to friends and colleagues, and animals that had killed people more by accident than design were targeted and shot. When these cases of 'manufactured' man-eaters were revealed and the unscrupulous official exposed, the state authorities issued an order that no official could issue a permit for shooting man-eaters or cattle killers in his own name. The spate of 'man-eating' stopped immediately.

# CHAPTER 4

# TIGERS OF THE SWAMP

By far the greatest number of fatal attacks by tigers take place in the Sundarbans (meaning 'beautiful forest'), a huge tract of swamp forest around the delta of the Ganges and Brahmaputra rivers in India and Bangladesh. Almost the entire Indian section is a formal nature reserve for the conservation of tigers, and there are three smaller reserves in Bangladesh.

Although nobody actually lives permanently in the centre of the 3,282 sq miles (8,500 sq km) area, there are 50 people killed by tigers every year. The people – about 35,000 of them – enter the forest to cut and collect wood, catch fish and gather honey. Some people go into the mangroves to collect *golpata* (nypa or mangrove palm leaves) and *hantal* (phoenix leaves). The fishermen work in the estuarine channels, catching fish and crabs, and, as tigers are good swimmers, the men are often attacked in their boats while sleeping or resting at night. Tigers have also been known to chase along the shore and intercept people passing in canoes.

The woodcutters tend to work in large parties, and operate at the edges of clearings. Only the honey collectors and phoenix leaf collectors walk deep into the mangrove forest on foot, and often as not they are so intent on finding the honey and beeswax that they are oblivious to tigers around them. When they spot a honey bee, they follow it intently back to the honeycomb. They are an easy target for tigers.

There are about 500–800 tigers, not only representing the largest population remaining in the world, but also one that is increasing at a rate of about 10 per cent a year. And they have learned that this itinerant population is easy prey. They will stalk their victim, sometimes for several days, until the right moment comes for them to strike.

Only 3 per cent of the Sundarbans tigers, however, were thought to be dedicated man-eaters, according to a study in 1971 by German biologist Dr Hubert Hendrichs, with another 30 per cent ready to attack on sight.

Hendrichs identified four categories of tiger operating in the Sundarbans: A-tigers that never attack people; B-tigers that attack if provoked, often only mauling a victim; C-tigers that attack on sight and may eat the body but do not return to a kill; and D-tigers that may search for human prey for days, often ignoring other prey, and will return to kills. D-tigers will even circle an area to which a dead body has been moved for several days.

Despite the few confirmed man-eaters, between 1969 and 1971, about 129 people were estimated to have been killed by them. When more reliable records were kept between 1975 and 1982, an annual breakdown of fatalities was available: in 1975 63 people were killed; in 1976 and 1977 the figure dropped to 40 and 37 respectively; during the next three years the casualty list climbed again with 48 killed in 1978, 52 in 1979, and 50 in 1980; in 1981 fatalities dropped to just 28, but then rose again in 1982 when there were 40 people killed by tigers. In 1988, the figure dropped to an all-time low of 11 deaths, but in 1992 fatalities went up again. The 40 deaths in that year are thought to be due in part to illegal entry by people, including children, who sought to take advantage of prawn harvesting. Indeed, in most years, over 80 per cent of the casualties are among fishermen who make up the largest working population of the three main professions – wood-cutters, honey gatherers and fishermen – represented in the

Sundarbans. More usually, though, the Sundarbans tigers (certainly Hendrichs's A and B tigers) enjoy a fairly catholic natural diet ranging from fish and crabs to water monitor lizards, deer and wild pigs. They swim frequently, moving from islet to islet across tidal creeks with a rise and fall of over 20ft (6m).

It is thought by local biologists and conservation officers that young tiger cubs learn that people are good to eat and easy to catch from their mothers, and so there is constantly a new generation which has mastered the practice of man-eating.

Hidden within the swamp forests, the Sundarbans' tigers have survived the wholesale killing of tigers that has gone on in the rest of the sub-continent. Their numbers have remained relatively constant whereas elsewhere they have dropped, bringing the tiger almost to the point of extinction in the wild. The swamp tigers, however, are still thriving.

The tigers usually approach their victims from behind, so at one time the woodcutters went about their business wearing a cheap plastic or rubberised face mask on the back of the head. The rear-facing, staring 'face' was thought to protect the wearer. From the tiger's point of view it looked as if the person was looking towards it, and therefore it could not take the intended target by surprise. The principle was based on the concept of false eye-spots used by certain butterflies and other creatures. The idea was introduced in 1986 at the suggestion of Arun Ram, a member of the science club in Calcutta. Some 2,500 out of a workforce of more than 8,000 honey gatherers, woodcutters and fisherman tried the scheme. During 1987, none of the men wearing the mask were killed, while 30 without were taken. Some men took their masks off when resting and were then killed. The mask-wearers reported seeing tigers following them, but they never attacked. Indeed, the scheme worked for a few years, but then the tigers figured out the scam and it is no longer so effective.

A problem for the authorities was the way tigers stalking people with masks almost always revealed their presence. This unnerved the honey collectors, and they handed in their permits to enter the area. In areas where masks were not being worn, the tigers took their victims completely by surprise, and although these people suffered more casualties, they were less frightened than the group with masks. 'Out of sight, out of mind' was clearly working here.

One myth was that the tiger always bites on the right side of the neck, and so the woodcutters and fishermen were advised to carry a stick over the right shoulder. Protective helmets and body armour were also issued but they were uncomfortable in the heat. Moreover, workers were supplied with fireworks to frighten away tigers.

In other attempts to discourage them from attacking humans, life-sized clay dummies dressed in the clothes of the local people have been wired up to 12-volt car batteries linked to transformers with a 240-volt output and placed in the forest or in fishing boats. The clothes have human scents, and the head of the dummy moves in the wind to give the illusion of a real person. If a tiger should attack, it receives a non-fatal but nonetheless painful electric shock and learns, perhaps, that people are not pleasant to catch.

The first tests were with six dummies, two dressed as fishermen, two as woodcutters and the remaining two as honey collectors. Each was clothed in the appropriate unwashed garb, to give the dummy an authentic smell. A fisherman dummy was placed next to the rear cabin opening in the sitting position in a country boat (a small boat with a cabin open at both ends and a curved roof of nypa or phoenix leaves) with the head facing away from the forest. Wire mesh across the opening was electrified. The tiger would have to push against it in order to get at the 'victim'. A fish basket containing freshly caught fish completed the deception. It was placed in position in April 1983, and almost immediately it was attacked by a tiger at night. The boat capsized as the tiger made its escape. In the morning the boat was salvaged. The electrified wire-mesh was ripped and the fuse had blown.

The woodcutter dummy was placed beside a tree at the high-tide mark in the act of chopping the tree down. The honey collector mannequin was placed as a man sitting on his haunches and positioned deeper into the forest. Both had wires looped across their torsos. The woodcutter was put in place in November 1983, but it was not attacked until February the following year. When examined, it had bite marks to the neck and claw marks on the thigh. The fuse had blown, indicating the tiger had received a shock.

The honey-collector was placed in a thicket in April 1984. A month later it had not been attacked, so the researchers made some adjustments that left more human odour on the dummy's clothes, and waited on a boat near by. Two nights later the researchers heard a loud roar from the forest. When they examined the dummy in the morning, they found it had been knocked over. The fuse had blown, and their were tiger pug-marks round the broken clay body.

During the first nine months, several of the dummies put out were attacked, and there were few human casualties during the experiment. Until its start, there had been 44 deaths a year on average due to tiger attacks, but in 1983 after the first tests began, the number of fatalities dropped. The dummies seemed to work, at least for a while. The year 1984 was renowned for its low attack record of tiger attacks. Following 40 fatal attacks in 1982, and 20 in 1983, those during 1984 were notable by their absence: there were just 12. The following year, 22 people were killed by tigers in West Bengal in the first three months: the killings were beginning to pick up again. In 1987 the death had risen to 24, but in 1988 just 11 people were killed. By 1989, the effect had worn off completely. Between January and May 50 people died, and in the first week of July alone, five woodcutters were killed by tigers and many more were attacked and mauled.

Since then, some key areas have been surrounded by electric fencing, and this has had some success in deterring tigers from entering. In other areas, pigs have been bred for release along the edge of the forest. The alternative source of food might prevent the tigers from taking people living there.

About 300,000 people live on islands at the edge of the Indian Sundarbans. Women and children face death daily from sharks and saltwater crocodiles as they wade through the shallows to catch shrimps. Even men in fishing boats are at risk. The Bengal tiger is a good swimmer. It waits in the dark shadows and attacks crews of boats anchored in the waterways. Times of greatest risk, according to Kalyan Chakrabarti, who was forest director for nine years, is early morning, late afternoon and about 11 o'clock at night.

Nevertheless, the local people are in awe of the tiger and consider it a great 'spirit of the forest'. They honour it with elaborate ceremonies, known as 'pujas'. Muslims and Hindus put aside their differences and worship at the same shrines at the edge of the forest. The Hindus believe the tiger is a manifestation of Banbibi – one of the forest spirits, and they worship her, while alongside in the same thatched-roofed shrine is Shah Jungli, the Muslim king of the jungle. Before honey collectors go to the forest in April and May, when the wild bees are building their honeycombs, they make offerings of leaves, flowers, sweets and sometimes coins to clay models of the spirits, but often as not they fail to placate them. There are many wives who have lost their husbands to tigers, the so called 'tiger-widows'. They receive 20,000 rupees ($520) compensation if their spouse was killed in a permitted area.

That tigers should live in the Sundarbans at all is unusual, for the entire area is without freshwater. Being a tidal delta, the water is brackish. Tigers drink large quantities of water, especially after a meal, and so one reason given for their man-eating habits is that they are forced to drink saline water, making them more aggressive. Hendrichs proposed this in his WWF report, suggesting that generation after generation of drinking saline water might cause liver and kidney abnormalities. Following the report, rainwater ponds were constructed. These are visited by tigers, but they still drink the saltwater as well. Indeed, Hendrichs also suggested that this habit of drinking salty water has given the tigers a taste for salt, and that they deliberately hunt down people because human flesh has a salty taste. However, other scientists, such as Choudhari and Sankhala, have noted that there are more tiger attacks in areas where the concentration of saltwater is lower.

Another reason, proposed by Indian conservationist Arjan Singh, is that the tigers are simply hungry. The delta was once home to large populations of wild buffalo, swamp deer, hog deer, barking deer and Javan rhino. Today wild pigs and spotted deer are all that remain. Few tracks at the freshwater ponds indicate that they too are low in numbers. The tiger then must rely on fish, amphibians and people for food, and will take anybody who happens to be passing as a useful protein supplement.

It could be, however, that some tigers are labelled 'man-eater' unreasonably. The honey-collecting season coincides with the tiger breeding season, so a man stumbling through the forest and not watching where he is going is quite likely to chance upon a sleeping tigress and her cubs. The mother's reaction is to defend her cubs, and the honey collector becomes a victim not of a man-eater but a 'caring' parent. After storms or typhoons, when food is scarce, tigers also raid fishing nets. If a fisherman should arrive to gather his catch before the tiger has finished, he is killed. Again, the tiger is not necessarily a man-eater, but it is defending its patch against an intruder.

Even if the bodies are eaten or partly eaten, the question begs whether the tiger is responsible. Crocodiles and wild boar are just two of the many scavengers present in the region that would have no hesitation in consuming a dead human body. And then there are the fishermen, woodcutters or honey collectors who owe substantial sums of money to money collectors. They conveniently disappear having been 'eaten by tigers' only to reappear after the moneylender has died.

Sankhala also draws attention to the known statistics and questions whether man-eating is as important even to the tigers that appear to indulge in it. He suggests that the meat value of 44 humans per year is insufficient to sustain just one tiger for more than two to three months, so people are unlikely to be regular prey of the Sundarbans tigers.

The attacks, however, occur to this day. At the beginning of January 1998, a young man, his father and two neighbours were crab-fishing in a creek when a tiger burst from the forest, jumped on to their canoe and attacked the young man. His father grabbed a stick and was able to push the beast back into the water before it could deliver its lethal bite to the neck. Nevertheless, the victim received three wounds from the tiger's sharp claws.

Man-eating, though, is not a new phenomenon in the Sundarbans. Tigers have been in the area and killing people for several hundred years. The diary of François Bernier, a French jeweller and adventurer, written some three hundred years ago, mentions tigers swimming out to fishing boats and grabbing members of the crew just as they do today. In 1869 a tigress was reported to have killed 129 people in the area. Indeed, the situation in the recent past was even more disturbing than it is today. During the six-year period from 1860 to 1866, 4,218 people were reported to have been killed by tigers in the Sundarbans.

Curiously, though, the tigers are rarely seen. Their distinctive pug-marks are commonly encountered, but the animals remain hidden. In fact, the Field Director of the Sundarbans Tiger Reserve has only seen tigers on eight occasions over 300 visits, and wildlife film makers Alan Macgregor and Dan Freeman, working on a film about the Sundarbans tigers for the BBC's *Natural World* series, spent several seasons tracking tigers but they were so elusive that the two film makers saw not one!

Outside of the Indian sub-continent, another trouble-spot is Sumatra, where people have been resettled after translocation from the rapidly diminishing forests of Java, Madura and Bali. Thirteen villagers were attacked in early 1980, with only two survivors to tell the tale. Four tigers were trapped and three killed in retaliation. Two villagers were eaten by tigers in December 1985. The cats were tracked and shot by police. Three plantation workers were killed in September 1989: unfamiliar with the habitat, they had fallen prey to tigers. The backlash, however, has had profound ecological effects. The new villagers have been killing the tigers – a very rare sub-species (*Panthera tigris sumatrae*) – and so the wild pig population has blossomed. The pigs, as a consequence, have become a major agricultural pest. Now the local farmers do not report sightings of tigers for the trophy hunting that results is disturbing the natural order of things.

In June 1997, there was another spate of attacks in Sumatra. Three people were killed in less than two weeks in the jungle around Bandar Lampung in the Way Kambas National Park, including a mother of two. Several forest workers were also mauled but survived, and as a consequence forest labourers planting mahogany and teak seedlings in a million-dollar forest replanting programme downed tools and refused to work until the tiger was caught. The following year, in March 1998, a Sumatran tiger thought to be responsible for the deaths of four plantation workers (one who had been killed by a tiger after a herd of elephants had trampled through his rice fields) was caught in a trap near the village of Fajar Bulan. It was tranquillised and taken to a zoo.

Elsewhere in the world, the conflict between people and tigers is endlessly repeated. In Thailand, for example, an old tiger with worn and broken teeth had taken to killing domestic dogs in and around the villages of Khao Yai National Park, about 130 miles (209km) east of Bangkok. One day, a young girl dropped her pencil between the floorboards of her hut, and crawling on all fours scrabbled about under the dwelling trying to find it. Little did she know that a tiger was hiding there. It grabbed her and killed her. A guard was posted to try to kill the beast, but he was killed and partly eaten too. It came back once too often, though: the next time it appeared it was shot.

On 15 January 1998, another tiger-attack incident occurred in the same park. A tiger that had previously received a gunshot wound to its foot and had taken to attacking people was shot dead after injuring two park officials. One of them was washing clothes behind his home at the edge of the forest when he heard the alarm calls of barking deer. Sensing danger, he high-tailed it to the house, but just as he reached the front door the tiger was upon him. It grabbed his right hand and tried to drag him into the woods. He managed to wrench it free, and shouted for help. The other park ranger came running but he too was attacked. The commotion alerted others in the building and they arrived with guns. They shot into the air and the tiger ran away, only to come back the following night looking for food. It roared and scratched at the door of a house in the village and its sole occupant made for the roof where he yelled for help. His father, a veteran forester who lived next door, appeared with his HK assault rifle and shot the tiger in front of the house.

The Malayan tiger (*Panthera tigris corbetti*) was hit badly by the demand for rubber which started in the 1920s and continued apace in the 1930s and 1940s. Vast tracks of its natural forest home were destroyed in favour of rubber plantations. Loss of habitat meant loss of prey, and plantations meant more people. Inevitably tigers took to hunting them. In the 1920s, the official records show that 9 people (7 men and 2 women) were attacked and killed by a tiger in a single day at near Jerantut, and another man-eater in the district of Besut, Terengganu, accounted for 36 victims before it was tracked down and shot.

In 1972, the tiger in Malaysia was protected by law and the population stabilised, but the increase in cattle, buffalo, sheep and goat farming has brought it into conflict with people. Although, tigers take mostly livestock there have been sporadic outbreaks of man-eating. During the course of a year in Pahang – Malaysia's largest state – there were four tiger attacks on people, two of which were fatal. Throughout the entire peninsula, during the 20 years from 1977 to 1997, however, there has been a total of only 12 incidents recorded. One occurred in January 1999, when a tiger pounced on a villager in an eastern Malaysia rubber plantation. The man was checking traps for wild boar when the cat rushed out from bushes. He grabbed for his hunting knife but it was too late: the tiger was upon him. The man fought back, and was bitten on the hands and the back of the head, but the tiger suddenly gave up the attack and disappeared into the forest.

During the Vietnam War, the tiger in the jungles of Vietnam made its presence felt to US troops stationed there. At first, tigers learned to associate the sound of gunfire with a super-abundance of food. They scavenged the corpses on battle-fields, and then took to raiding the many war graves of their dead bodies. Finally, they turned to man-hunting. One marine was grabbed by the leg while lying in his fox-hole near Da Yang, and was dragged by the tiger into the jungle. Fortunately he became entangled in barbed wire, and his calls brought the rest of his platoon to drive the tiger away.

In the Russian Far East, an estimated 200 Siberian tigers (*Panthera tigris altaica*) have been attacking livestock and, by the summer of 1985 in some regions, had become a threat to people living there. Tiger-watcher Victor Zhivotchenko has come across reports of two male tigers attacking a forest worker and a hunter attending traps in the Sikhote Alin Reserve. This is not new. Russian traveller Nikolai Przewalski (who discovered a wild camel and a breed of wild Eurasian horse in Mongolia) wrote about many attacks on people by tigers over a century ago. In the early part of the 20th century, Cossack troops had to be brought in to guard the Chinese workers who were building the Trans-Baikal Railway. Siberian tigers were taking workers in a series of events reminiscent of the Tsavo lions (see p.67). In Manchuria, weekend markets had to be closed because tigers were on the rampage. The attacks continue to the present day.

In the Primorye territory, where the temperature drops to -40 degrees C in winter and snow can be 20in (50cm) deep, tigers take to following human paths and tracks. It is a big animal and sinks into the snow; far better, then, for it to follow people. The result is many encounters between tigers and people every year; sometimes they have been fatal. In February 1976, for example, a tractor driver was killed and partly eaten in the Lazovsky area. It was the first case in 50 years. In 1981, a resident of the Terneysky district was consumed. And in the mid-1980s, young tigers had taken to attacking people, although many were not fatal.

In December 1997, a Siberian tiger took to man-eating near the village of Sobolinoe in the north of Primorye. The first hint that it was on the loose was when a hunter failed to return home after an excursion into the taiga. Villagers scoured the area and discovered tiger footprints near a forest hut. They also found scratch marks on a nearby outdoor lavatory, but there was no body. A few days later, another villager was setting traps in the forest when he was attacked. His family had pleaded with him not to go alone, but he ignored them. His father and neighbours found two fingers, part of his scalp, and a chain attached to a watch. The authorities brought in trackers from TIGER, an organisation that is dedicated to conserving tigers but charged with tracking down man-eaters. They followed the tiger for two days and eventually shot it.

Examination of the body revealed that the tiger had been shot previously in the leg. Maybe its injury had prevented it from catching its natural prey, and it took to eating anything that it chanced upon, like a hunter or two bending over when setting traps.

Elsewhere in the 1980s, tigers were leaving their forest homes and entering built-up areas in search of food. In 1986, a tiger was shot near a crowded trolley-bus stop on the outskirts of Vladivostok. In 1987, in Nakhodka, a soldier opened the curtains of his suburban house and, through the frosted pane, came face to face with a Siberian tiger. And commuters heading by train for the main container terminal in the town were held up by a tiger on the tracks. Several attacks on people and livestock were reported.

Many of the Siberian tiger attacks have been by young individuals which have been chased away by their mothers and must fend for themselves. With an increase in agriculture and logging, a continuation in the hunting of wild boar – the tiger's prey – and a successful conservation programme resulting in a larger number of tigers surviving, natural prey is insufficient to sustain the population. They turn to livestock and people instead.

That the tiger is a potential man-eater wherever it may live, there is no doubt, but it seems that human behaviour rather than tiger behaviour – as is so often in these cases – is frequently the trigger for incidents of man-eating. The distinguished cat specialist from the Max Planck Institute, Professor Dr Paul Leyhausen, summarises the general feelings of the scientific community:

'Certainly every tiger is a man-eater, in the same sense that every car-driver is a potential man-killer – if he wants to, or is irresponsible, he may kill a human being. If a tiger wants to, he may kill and even eat a human being. However, the fact is that tigers rarely do so. Normally it is quite possible to walk unarmed through tiger-inhabited country with no fear of being attacked. Furthermore, in the majority of cases, even when a tiger becomes a man-eater, this is mainly because human carelessness gives him an opportunity. Just as there are numerous people who will cross a busy road without first looking left and right, there are people living all the time in the neighbourhood of tigers who are rather too careless when there has not been an accident for years. For obvious reasons we do not abolish cars, and for perhaps less obvious, but far more important reasons, we must not abolish the tiger.'

# CHAPTER 5

# MAN-EATING LIONS

Lions (*Panthera leo*) are big and powerful cats, and their tendency to hunt co-operatively in a pride makes them formidable predators, well able to pull down animals much bigger than themselves. In the wild, they rarely grow larger than 9ft (2.7m) long and weigh 550lb (249kg), although there was a huge lion shot near Hectorspruit in the eastern Transvaal of South Africa in 1936 that scaled 690lb (313kg). While game, such as antelope, wild pigs and buffalo, makes up most of a lion's diet, occasionally people are taken as well.

One of the earliest records of a man-eating lion appears to be an Assyrian ivory panel from Nimrud, dating from the 8th century BC. It depicts a man being grabbed by the throat by a lion. The panel is exhibited at the British Museum in London.

One of the most successful group of man-eating lions must have been a pride of 17 lions that lived in a game reserve at the northern end of Lake Nyasa in Tanzania. The pride and its offspring went on a 15-year rampage that accounted for about 1,000 to 1,500 human deaths. Records were not kept so nobody is certain of the final toll. The pride was known as the 'Njombe Man-eaters'.

In 1925, two lions were reported to have devoured 124

people in Ankole in Uganda, and another pride, consisting of five sub-adult lions, accounted for 20 human deaths in the Lindi district during 1935. Previously in the same area, lions killed and ate 140 people in the course of a few months.

During the Second World War, reports emerged describing how 153 people were eaten by two prides of lions – one group of five and the other of eight – in the sub-chiefdom of Wangingombe, in Uganda.

These bouts of man-eating seemed to coincide either with a reduction in the numbers of game animals or with inclement weather. In 1904, for example, man-eating lions were prevalent in Rutshuru in eastern Congo, their newly acquired eating habits coinciding with a marked reduction in the herds of topi and kob.

In the main, though, humans seem to fall foul of solitary or 'exiled' lions without a pride and old or infirm individuals who are incapable of running down a meal. In 1977 a rogue lion, which attacked twelve people and killed eight in Tanzania, was found to have only three legs.

People, particularly anybody walking alone in the bush, present an easy feeding option for lions, that is, after the initial fear of man is overcome. But once a lion has tasted human flesh it seems to seek more. There is, for example, the case of a rogue lion taking 14 people in Malawi during the course of just one month. Another in the Numgari District of what was Portuguese West Africa accounted for 22 local village folk in eight weeks. It was shot in September 1938. And in October 1943, a rogue lion was shot after it had killed and ate 40 people in Kasama District of what was Northern Rhodesia.

At the Lake Manyara National Park in Tanzania, a lion took an interest in a village rubbish dump where it came into contact with human smells, and lost its fear of people. One night a drunken man staggered about in the road and the lion grabbed and ate him. Having acquired the taste of human flesh, it despatched another two villagers before it was isolated and killed.

Further to the south, in South Africa, the stories are legend. The Landdrost, Joseph Sterreberg Kupt, for example, was looking to buy some oxen for the Dutch East India Company and arrived at a kraal and pitched camp. During the night the horses and cattle were restless, and the entire camp left their tents, guns in hand, to investigate the disturbance. About 100 yards (100m) from the tents they could see a lion standing. When it saw the people assembling, the lion moved about 30 yards (30m) further on. It stopped behind a small thorn bush at the bottom of a steep hill where Kupt thought it had hidden a young ox. The men fired about 60 shots, hoping to kill the beast. The moon shone brightly and so they could see clearly that there was no movement from the vicinity of the bush.

Kupt's people settled the cattle, but then noticed that a sentry – Jan Smit from Antwerp – was missing. They called for him, but there was no reply. They then thought that the body the lion had behind the bush was that of Jan Smit. A group of men moved cautiously towards the bush, but the lion was still alive and stood up and roared. They all retreated rapidly, but not before they spotted the cocked musket of the sentry. The men fired another barrage of a hundred shots at the bush. All was still, and so a marksman worked his way forward. Again, the lion roared and leapt at the man. He threw a thunderflash, and the others fired a fusillade of ten more shots. The lion hid again behind the bush. As dawn came up, the lion climbed the hill with the sentry's body in his mouth. Another 40 shots were loosed, but the lion escaped with his victim. In order to retrieve the body 50 armed Hottentots were despatched. They found the lion a few miles further on. It began to run away but they gave chase. The lion then turned and ran straight towards its pursuers. One of the group was attacked, but the head of the kraal put himself between the lion and the man, and stabbed it with an assegai. The rest of the Hottentots drew their weapons and speared the lion until it resembled a porcupine, but the lion was still full of life. Eventually, the marksman put a ball in the lion's eye which caused him to fall. Others shot him dead. The lion was recognised as one which had taken one of the Hottentots from the kraal and eaten him only a few days before Kupt's visit.

Kupt's experience was one thing, but an altogether more extraordinary story was that of the 'man-eaters of Tsavo'. It all began in 1896, when the British in British East Africa started to build a railway from Mombasa on the coast to Kampala in Uganda. It was officially called the Uganda Line, but became known as the 'Lunatic Line' for it was said along the 580 miles (933km) of track across the savannah, many large rivers and the Great Rift Valley, went 'from nowhere to utterly nowhere'. In reality it was to be the means by which Europeans could journey into the interior of Africa, a trek which until the construction of the railway had to be made on four-legged transport or on foot. It was also a means by which it was thought the slave trade could be controlled or even stopped. A caravan route ran for 2,000 years from the interior to the coast. The trade was in ivory and slaves, and Mombasa was a slave port. The railway was thought to be a way to change all that. Unfortunately, the local people were in fear of it and local witchdoctors prevented them from working on it, and so most of the labour for the enterprise came from India. The 'coolies' as they were known were drafted in from one part of the British Empire to another, but over 90 per cent succumbed to disease. Of the rest, 140 were to die a horrible death. It was to become one of the most devastating man-eating stories on record.

Drama struck in March 1898, during the construction of a 100 yard (100m) long railway bridge over the Tsavo river in Kenya about 132 miles (212km) from Mombasa. In charge was Lt Colonel John Henry Patterson, who had come from engineering projects in India. Hundreds of men were involved in the building of the bridge and the railway, and so there were many work camps scattered for 30 miles (48km) along the railway tracks.

Things began to go wrong shortly after Patterson's arrival. He started to receive reports that workers were disappearing and that a lion was to blame. Patterson refused to believe it at first but soon the reality dawned: there was not one lion but two and they were targeting workers from the railway. Locals called them 'the ghost' and 'the darkness', and thought they were devils in the shape of lions.

The first of his party to disappear was a powerfully built Sikh named Ungan Singh. He was one of several workmen resting in a large tent, and was sleeping closest to the entrance. The lion pushed its head through the flap and grabbed Singh by the throat. He yelled 'Let go' and flung his arms around the lion's neck. The others in the tent were petrified. They saw the man dragged away in an instant. In the morning, Patterson saw the pug-marks around the tent and the furrows in the sand where the victim's heels had dragged across the ground.

Hunting parties were sent out and, guided by pools of blood where the lion had stopped to eat (lions have a habit of licking off the skin of their prey so as to get at the fresh blood below), they soon found the remains of the man. The ground was covered with blood and scraps of flesh, skin and bones, but the head, which lay with staring eyes a short distance away, was still intact. There and then, Patterson vowed to rid the area of its man-eating lions, but on this occasion, the animals evaded the trackers.

With so many camps from which to choose, the lions had a field day. As Patterson waited in a tree, rifle and shotgun in hand, at one camp, the lions attacked people at another about a half a mile away. First the lions roared, then silence, followed by the shrieks and cries as another victim was seized. Every time he staked out a camp, the lions went about their grisly business elsewhere. Thorn corrals or 'bomas' were erected around the camps, and fires were kept burning all night, but the lions ignored them. Used to stalking and hunting game through the thick Tsavo bush, they broke into the compounds with ease and dragged their victims out through the thorn fence and into the bush.

Some workers were lucky to be alive, and their encounters brought some amusement into the camps which were by now very tense. On one night an Indian trader on a donkey was attacked but the lion's claws became caught on a rope attached to some oil cans. The clatter frightened the lion and it ran back into the bush, the cans clanging behind it. The donkey rider was unhurt. On another occasion, a lion burst into a tent but its claws caught on the mattress of his intended victim, a Greek contractor; it too made a hasty retreat, together with the remains of the bed. In the meantime, the other lion broke into a tent shared by 14 coolies. It mauled one man, but in the confusion picked up a sack of rice and made off with it. It dropped it 'in disgust', as Patterson wrote later.

By this time, the railhead had moved further to the north, leaving 300 or so workers to build the bridge. The lions focused all their attention here. To all accounts they were strange-looking beasts. They were both males, but they had very little hair in their manes, a trait shared with other lions at Tsavo and thought to be a genetic peculiarity of males in the area. Indeed, there are two types of male lions – some with rather scraggy manes and others with no manes at all. The local oddity seems to be an adaptation to hunting among the thorny thicket that is so characteristic of Tsavo.

Why they took to attacking humans in the first place is unclear. Some believe that Tsavo lions are prone to man-eating, an attribute they possess even to this day. The habitat in which they live is tough and unforgiving; it is predisposed to long droughts and food can be difficult to find. Many Tsavo lions tend to hunt in pairs and intercept anything that they may chance upon during the night. People would be seen no differently than their natural prey. Evidence suggests that at the time Patterson was building the railway, the area was in the grip of a drought, and the lions would have been alert to any feeding opportunity, especially an entire work-force unused to the ways of Africa.

There are other explanations. Slave caravans passed through the area and many slaves died or were dying. They were left where they dropped along the route. Lions and hyenas would have taken advantage of the bodies. Maybe man-eating had its roots in this miserable business.

The 1890s was also a decade when an outbreak of rinder-pest, a serious disease of hoofed animals, took its toll of wild antelope, zebras, buffalo, wildebeest and gazelles – all primary foods of the African lion. Deprived of their natural food, the Tsavo lions might have turned to people; after all, they were very abundant at the time. Not only that – many coolies died of malaria during the construction of the railroad. Burial was haphazard. Dead bodies were left with a burning coal in the mouth, in the belief that the corpse would self-cremate. Lions – able scavengers as well as noble predators – would have had easy access to this bonanza of carrion. It would have been an easy move from the dead to the living.

At the Tsavo bridge camp things became very uncomfortable. Homing in on the hospital tent, the lions caused havoc. One broke into the tent, mauled two patients, and carried away a third. The tent was moved and surrounded by a thick thorn barricade, only to receive a second attack when the hospital's water carrier was killed and dragged away through the fence and into the bush. Witnesses told how the man had been seized by the foot, and in desperation had tried to prevent his abduction first by grasping a large heavy box and then a guy rope which broke. When the lion pulled him free from the tent, it went for his throat and suffocated him with its powerful grip. His skull, jaws, a few bones and the palm of one hand with a couple of fingers attached were found later.

The tent was moved yet again and surrounded by an even thicker fence. In the meantime, Patterson and the hospital doctor hid in an open goods wagon which was placed in a siding near the old hospital enclosure. Cattle were put in as bait. The two men remained awake all night and, armed with rifles, were ready to ambush the lions. Sure enough, one of the lions appeared, but it ignored the decoy and began to stalk the two men instead. Without a sound, the lion charged them and, suddenly realising their dilemma, the two fired simultaneously. The surprised lion took off and the two man-eaters were not seen for some time.

For a couple of months the lions failed to visit the Tsavo Bridge camp, although people were disappearing from camps elsewhere along the line. Patterson used the time to take stock of the situation. He set a trap by building a two-chambered box of wooden railway sleepers, tram-rails, telegraph wire and heavy chain. The two chambers were divided by the heavy rails set just 3in (7.6cm) apart. In one chamber sat men with rifles, hoping to entice the lions into the other. On the lions' side a trap-door was fastened to chains and wire, which in turn were linked to a spring on the floor of the cage. As a lion leapt into the cage, it would depress the spring and the trap-door would drop behind it. To embellish the deception, a tent was pitched over the trap and surrounded by a thorn fence. The lions, though, failed to appear.

71

Elsewhere along the railway, however, the lions were active. Four men were eaten and one so badly mauled that he died later. At Tsavo, where the lions had stayed away for a while, the workers became complacent. Several slept outside their tents although they remained inside their *boma*. But just as suddenly as they had disappeared, the two 'demons' reappeared. One leapt into a *boma* and, despite being shot at and having flaming torches thrown at it, it grabbed a coolie and dragged him through the thorn fence. It had become so daring that the other lion joined it just 30 yards (30m) outside the *boma* and they both finished their meal, taking no notice of the shots being fired at them.

Each evening for weeks thereafter the lions followed a particular routine. At dusk they would roar defiantly and when they grew silent, the hunt would begin. They visited camps up and down the line. The coolies whispered to each other, 'Beware brothers, the devil is coming.' A scream in the night would betray one less person at work the following morning.

Then the lions changed their tactics. Until this point, one lion would enter the *boma* while the other waited outside. In future, the two hunted together. In a camp near Tsavo station they made their presence felt. From an iron hut nearby, the permanent way inspector heard the commotion and took his rifle and shot into the darkness. In the morning, he and Patterson followed the trail. They were stopped in their tracks by growling, and pushing slowly forward through the bush, they found the remains of that night's feast. The workman's legs, an arm and half the torso had already been consumed.

Eventually the coolies had had enough. They downed tools, hijacked the next train to the coast and left. The railway bridge construction ground to a halt. Those who remained took to sleeping in trees, on wooden shelters on the top of high water tanks, or in specially dug trenches protected by logs underneath their tents. Many opted for the trees. In fact, so many ended up in one particular tree that, one night when the lions paid a visit, the trunk cracked and the tree fell, spilling its human congregation on to the ground. The lions ignored them; they had already caught their food for the night. Others in the camp could hear the crunching of the bones and the satisfied snarls of the lions.

Meanwhile, assistance was requested from and agreed with the District Officer, Mr Whitehead, who decided to take a look at the situation for himself and set off for Tsavo. Patterson despatched his boy to meet him at Tsavo station. The boy returned trembling. There was no train and no soldiers. But a large lion stood defiantly on the platform. The station staff had locked themselves in the station buildings. With no sign of the train, Patterson assumed they would arrive the following day and went to dine alone.

The District Officer's train was, indeed, late. It arrived later in the evening. But, walking from the station to Patterson's camp in the dark, Whitehead and his servant were intercepted by one of the lions. It leapt down from a bank, and landed on Whitehead's back. Whitehead fired his carbine and the startled lion jumped off; but it grabbed his servant and dragged him away. Whitehead had bloody claw marks down his back as a result. The servant was eaten.

On the same day, the Superintendent of Police arrived with 20 armed men who were deployed to hunt down the cats. That very night, however, one of the lions entered the wooden trap and was caught as the trap-door dropped. It made such a commotion that the armed guards lying in wait were petrified. Even though they could have touched the lion with their guns they somehow failed to hit it with a single shot. Instead, a wild shot hit one of the bars on the door and the lion escaped. Patterson and the police tried to track the beast, but failed to find it. They searched for two days, but without success. The police were withdrawn and Patterson was once more on his own.

On 9 December, one of the local people came running into Patterson's camp. 'Simba, simba,' he yelled (Swahili for lion). One of the Tsavo lions had tried to grab a man from a donkey but he escaped; they were now enjoying a meal of donkey flesh, starting on the hind quarters, as is the custom of lions, and working towards the head. Patterson and his informer crept slowly towards the two diners, but the man stepped on a dried twig. The sound alerted the lions and they ran off into a dense thicket.

Gathering a large group of coolies together, Patterson decided he would flush the lion out into the open. They surrounded the thicket and, shouting and banging on pots and pans, made as much noise as they could. The lion emerged, Patterson raised his heavy-duty, double-barrel rifle and ... 'click' ... the rifle failed to fire. The lion raced past a surprised Patterson who, remembering he had a second chance with the other barrel, let off another shot. It hit the lion in the back but without faltering it continued to run off.

Then Patterson had a brainwave. He got his coolies to build a rather rickety, 12ft (3.7m) high shooting platform or *machon* on wooden stilts and, tying the remains of the dead donkey to a tree stump, sat out the night waiting for the lion to return to its kill. With his normal gun-bearer coughing badly and likely to spoil the ambush, Patterson undertook the vigil alone. It was not long before he heard a twig snap and a sigh. The lion was out in the darkness and it was circling the platform. It had ignored the donkey and was stalking Patterson himself. One leap and the entire edifice would have toppled. As the tension rose, something hit Patterson's head: an owl had mistaken him for a tree. Recovering his composure, he raised his rifle and fired a shot into the dark, in the direction he had heard a growl. The lion roared loudly. It was hiding in the thick bush. Patterson fired repeatedly, and the growling subsided and finally stopped. He had killed one of the Tsavo man-eaters. In the morning, the workers retrieved the lion's body. It was a large maneless beast, 9ft 8 in (2.9m) from the nose to the tip of the tail and standing 3ft 9in (1.1m) high. It required eight coolies to carry it back to the camp where it was skinned.

The other lion was not seen or heard for several days, but eventually a railroad inspector heard noises outside his house. Thinking it was a drunken coolie, he did not go outside, which was just as well: the intruder was the second Tsavo man-eater. Failing to find human flesh, the lion made do with a couple of the inspector's goats. On the following night Patterson staked out the inspector's house. He tied three goats to a 250lb (113kg) section of railway rail and waited in a nearby hut. Just before dawn the lion appeared. It grabbed one of the goats and began to carry it off, along with the heavy rail and two live goats. Patterson ran out, fired at the lion but hit one of the goats instead. The lion made off with its strange entourage. At first light, Patterson and several men tracked the lion. The trail was easy to follow, and they quickly found the beast feasting on one of the goats. As the men approached, the startled lion looked up and charged. It ran straight past its pursuers and escaped into the bush.

With the success of the previous lion kill in mind, Patterson had another platform built – a little sturdier this time – and the goat carcasses were left out as bait. This time Patterson was not alone. He and his gun bearer took turns to watch, and right on cue, the lion reappeared. Patterson fired both barrels of his double smooth-bore, hitting the lion in the shoulder, but it ran off. This time it was bleeding. The hunters followed the trail of blood, but it grew faint. They had lost the second lion yet again.

The railway- and bridge-building gangs hoped the lion had slunk away into the bush and quietly died, but they were wrong. On the evening of 27 December, it visited Patterson's camp. It went from tent to tent, but fortunately the camp was deserted. The lion continued its search for food, and found some men sleeping in a tree. They were out of reach. The men shouted, alerting Patterson in his tent nearby, and he came with his rifle. It was pitch dark and all he was able to do was shoot into the blackness. The lion made off.

On the next night, Patterson and his gun bearer climbed into the same tree – narrowly avoiding a poisonous snake – and waited for the lion to return. At 3 o'clock in the morning, while the gun bearer was on sentry duty and Patterson was asleep, the lion came back. Sensing something was amiss, Patterson awoke. The lion was stalking him once again. The two watched it close in on them and, when it was about 20 yards (20m) from the tree, Patterson fired his .303 rifle and scored a direct hit. The lion was still standing, though. It growled loudly, and ran away. Three more shots rang out, one hitting the retreating lion.

At dawn, the two men tracked the lion. There was more

blood this time, and they found the injured animal about a quarter of a mile away hiding in the grass. It snarled aggressively and bared its teeth. Patterson took his time and aimed his rifle carefully. He fired, but the lion charged forward. Patterson shot again, and the lion fell. But seconds later, it was up again and continuing its charge. A third shot rang out but the lion just kept on running. It was like a bad dream. Patterson reached behind to his gun bearer for his Martini carbine, but he was no longer there. The man had wisely shinned up the nearest tree. Out in the open without a weapon, Patterson quickly followed suit. Fortunately for him, one of the shots had shattered one of the lion's hind legs. Had it not, Patterson would never have made it to the tree. But, safely ensconced on an overhanging bough, Patterson took the Martini and shot the snarling beast. It collapsed. He leapt down to take a look, but his excitement was premature. The lion reared up and charged yet again. Patterson shot it in the chest and in the head. It dropped to the ground, just 5 yards (5m) from Patterson, biting defiantly at a fallen branch. Finally, it was dead. The railway workers were ecstatic, and it was all Patterson could do to stop them from tearing the animal limb from limb. They carried it back to camp, and found that it had been hit six times. It measured 9ft 6in. (2.9m). It too was skinned.

Patterson, known now as the 'devil killer', had the skins mounted as rugs, and on 30 January 1899 he received a silver bowl from the coolies in gratitude for his bravery. The bridge was finished, and Colonel Patterson left Africa in 1899, only to return again in 1906 when he spent several years there as a game warden. He wrote his famous book *The Man-eaters of Tsavo* the following year. Many years later, two films were made of Patterson's adventures – a 3-D movie in 1952 called *Bwana Devil*, created by Arch Oboler, and William Goldman's feature called *The Ghost and the Darkness*, released in 1986.

In 1925, Patterson sold the lion skins to the Field Museum of Natural History in Chicago for the sum of $5,000, and the rugs were converted into life-size mounts. They went on display in 1928, and they can be seen there to this day.

Twenty-eight Indian coolies lost their life to the man-eaters of Tsavo, together with countless local people for whom no official records were kept.

Patterson's man-eating tales of the late 19th century however were not limited to Tsavo. Shortly before he left Africa he was on inspection duty to Voi when a stretcher with a European road construction engineer, Mr O'Hara, was brought into camp. Not far behind followed the man's wife and children. Patterson met them and heard their horrific story.

The family had been asleep in a tent when Mrs O'Hara thought she heard a lion. She awoke her husband, and he went outside with his gun. He asked the sentry if he had seen anything, but he replied, 'Only a donkey.' O'Hara returned to the tent. Later in the night Mrs O'Hara awoke with a start. She felt as if the pillow was being pulled from beneath her head. She found her husband had gone. Going outside, she found him lying between some boxes. She shouted for the sentry to help her, but he refused, saying that there was a lion beside her. Mrs O'Hara looked up to see the animal looking down at her. The sentry fired and the lion ran off. Mr O'Hara was dead.

Several workers lifted the body back into the tent, but the lion returned for its prey. They fired again and it ran away, only to return constantly during the night. Mrs O'Hara took to firing intermittently through the fabric of the tent to keep it at bay. Eventually morning came and the lion left them alone. The body was brought to the Voi hospital. The lion was killed later by a local hunter using a poisoned arrow.

Patterson was also to chance upon another series of disturbing events. A man-eating lion was terrorising railway staff at Kimaa. One night it actually leapt on to the station building and tried to tear away the corrugated roof to get at the people inside. One of the men sent a telegraph message with the immortal words, 'Lion fighting with station. Send urgent succour.'

The Superintendent of Police, Mr Ryall, arrived with two colleagues, and their carriage was put in a siding at Kimaa. The three intended to shoot the miscreant, but failed to find it during their first inspection of the immediate area. They returned to the carriage, had dinner, and then stood guard. All they saw in the darkness were two glow-worms – at least, that is what they thought. The pinpoints of light were reflections of their own lights in the eyes of the lion, which was watching them intently. Ryall remained alert while the other two slept. But unfortunately he dozed off too. The lion approached silently. The sliding door on the carriage was not quite shut and the lion was able to nudge it open and pushed its way inside. The tilt of the carriage, however, caused the door to close behind it and the lock snapped shut. The lion was trapped in the carriage with the three men. It leapt immediately on Ryall, planting its hind feet on one of the other men sleeping on the floor. The third man looked down in horror. His only escape was over the lion's back and into the servants' quarter. Somehow he vaulted over the snarling beast and made for the connecting door. The coolies on the other side had pulled it shut, however, and were scared to open it. He pulled with all his strength and managed to open it sufficiently to squeeze to safety.

In the meantime, there was a loud crash, and the lion broke through one of the windows, dragging Ryall with him. The man on the floor leapt out of the opposite window and fled to one of the station buildings. Ryall's remains were found the next morning about a quarter of a mile away. Some time later the lion was trapped, kept on view for a few days and then shot.

Many years later, another maneless lion took to man-eating near Mfuwe in Zambia. During the course of two months in 1991 it attacked and ate at least six people. The authorities went after what they thought were the culprits and shot several lionesses, but they had not realised that the man-eater was the lone maneless male.

Its last victim was an old lady in the village of Ngozo. California big-game hunter Wayne Hosek, who happened to be nearby, went to investigate. Together with a safari guide and a couple of trackers, he scoured the area and discovered from the local villagers that the lion had later returned to the old woman's house and carried off a bag of clothes. The following night it was seen to 'play' with its prize beside the Laungwa river, not far from the village. Hosek felt that this was the place where they could set up an ambush, so they erected a blind and waited. They sat there for a week before the lion appeared. In late August, it was shot.

The 'Man-eater of Mfuwe', as it became known, was another large beast, in fact, the largest known man-eating lion. It measured 10ft 6in. (3.2m) nose-to-tail, 5ft (1.5m) from floor to ear tips, and weighed 500 lb (227kg). Its stuffed body was presented to the Field Museum.

Lions have continued the Tsavo tradition elsewhere in Africa. In August 1984, an 18-year-old German girl, camping with friends in northern Botswana's Okovango swamps, was pulled from her tent and eaten by a lion, while in August 1989, reports were circulating of a small pride of man-eating lions that had killed and eaten nine people near Kondoa, in north-central Tanzania. Three were shot, but one was still at large at the end of the year. In November 1996, a man thought to have been drunk, walked through a wildlife park outside Johannesburg and became dinner for five lions. All that was left, according to rangers, was a pile of bloody bones.

In November 1998, a six-month-old lioness at the Camorhi game farm in South Africa's Free State found a gap under the fence of its enclosure and grabbed a two-year-old boy. The child's grandfather rushed over and held the lioness by the nose, twisting it in order to make the animal let go. A farm ranger eventually jumped on the lioness and forced it to release the youngster, who was rushed to hospital. He had to have 60 stitches in his right calf. The following year in May 1999, a group of nine tourists watched in horror as their guide at the Okapuka game farm outside Namibia's capital Windhoek was grabbed and eaten by a lion. The man had been trying to free a mechanism that was used to feed the lions when he was attacked.

More recently, in August 1999, a 19-year-old British tour assistant camping in Matusadona Park in Zimbabwe was attacked in the night by a pride of about 12 lions. At approximately 1.30 in the morning an armed safari guide was woken by the young man's calls, and in the moonlight saw him running away from his tent. The guide grabbed his gun, but was unable to shoot for fear of hitting the man. Instead, he ignited flares and called for a vehicle to drive at the lions. By the time he reached him, it was too late. An ageing lioness was thought to have led the pride. She was tracked and shot dead.

At the other end of the country, another lion crisis had been developing. In 1987, a pair of delinquent males accounted for about 30 people, including the local game warden, in the Tunduru district of southern Tanzania. In January, they mauled to death a man and woman and killed a woman and her baby in a residential suburb of the town of Tunduru, about 344 miles (550km) south of Dar es Salaam. The district is on the border with Mozambique, from whence the lions apparently arrived. These lions are thought to have acquired their pre-dilection for human flesh from the casualties of Mozambique's bush war. The problem was then exacerbated by the poaching of game that deprived the lions of their normal food supply. The local people had few cattle or any other domestic animals, and so relied on the game for meat. They decimated the area. As a result, the lions targeted homesteads and compounds, scratching at doors and windows and tearing off roofs just as the Tsavo pair had done. One lion chased a dog into a hut, but once inside it seized a 10-year-old boy and ate him. The lion remained in the house all night while the rest of the family took refuge in another room. In the morning it left, but was tracked down by the boy's father and shot. A second lion tried to take two of the hunting party and was also killed.

About 250 lions were killed in order to rid the area of its troubles, but how many of the casualties were man-eaters is unknown. Scientists working in the area suspect that many deaths were the work of paid assassins or 'lion men', who murdered people on contract. The success of these murders relied on rampant superstition in local communities, and the most horrific lion-men killings took place in what was the Central Province of Tanganyika – present-day Tanzania. The phenomenon was that of lycanthropy – a form of black magic in which people take the shape of animals. In Tanzania, the African equivalent of the werewolf was watuSimba, the were-lion.

It was said that if a dead man's body remained flaccid, rather than stiffening by *rigor mortis*, his grave should be watched carefully. After three or four days, he would dig his way out and, although weak at first, he would gradually gain strength before embarking on a bout of man-eating. The story has a striking similarity to that of the European vampire. The lion-men, however, were altogether more sinister.

In 1920, 200 people succumbed to watuSimba. Their bodies were found with claw marks, but curiously they looked as if they had been despatched with knives. Lion hunter W. Hitchens described in *Wide World* how he discovered what was behind the mystery. On one occasion he thought he was tracking a lion, but then came across a youth wearing a lion skin. The unfortunate boy was drugged with hashish, wore gloves equipped with a lion's claws and carried a long and lethal-looking stabbing knife. It was all part of a deception to extract money from the local people in a particularly obnoxious protection racket. Secret lion-men societies were responsible for ritual murders that were made to look like lion kills.

In the Usare regions, where the boy had been caught, it was estimated that there had been at least 200 fake lion killings. The boy and his employers were eventually convicted of the murders. The racket was run by witchdoctors. In areas where genuine man-eating lions were operating, they offered protection, for a price. It also enabled adversaries to settle old scores. The watuSimba or Mboji was hired to assassinate opponents for 40 shillings a killing, and then their relatives were offered the same services in revenge. It was a profitable fraud, but several of the lion-men were eventually caught and hanged, and the business went into decline.

In 1946 it had a revival, but had moved further to the south near the village of Singida where 30 men were murdered. One woman escaped her assassins and described being attacked by youths in lion skins. Such was the strength of the local mafia, she later retracted her story, and claimed a real lion had mauled her. In January 1947, the ring-leaders were rounded up and taken to the Dodoma court but, because the drug-dependent lion-men remained in the field, the killings did not stop. Also, real lions were enlisted. Young lions were captured and their teeth filed to vicious points. The death toll that year came to 103, and it was not until many of the leaders had been hanged that the killings ceased.

The real horror of the watuSimba cult, however, became evident as children, both boys and girls, were kidnapped or sold to the witchdoctors. They were kept in the dark in places where they could not stand upright and developed a crouching walk. Their wrists were broken and their hands tied back to simulate the foot-pads of big cats, and the tendons in the leg were cut to give them the right gait. They were fed on meat and lived like animals. Victims were sometimes consumed. The court heard how a five-year-old girl had been taken from her mother by a lion-man, and her remains, such as they were – her skull, teeth and few other parts – were found later. The witchdoctors responsible for the killing were caught and hanged, but the lion-man who carried out the atrocity was never apprehended.

During the 1950s, the killings started again; indeed, there was a court case in 1957. From time to time their activities come to light to this day.

On Mozambique's border with South Africa, however, real lions are still feasting on people today. In short, the animals have taken to eating those who are evading the law. In the Kruger National Park of South Africa illegal immigrants from Mozambique who seek work in Johannesburg have been taking a short-cut through the park. Lying in wait for them have been the park's lions. Nobody knows how many people have been killed, but the problem was noticed in 1980 and has steadily become worse. At first only one or two people were known to have been taken by lions each year, such as the mother and her son killed close to Pafuri Camp on 31 December 1996, but in 1997 11 cases came to light; a man and a woman, for example, were killed and eaten by lions about 19 miles (30km) from Parfuri in July of that year. According to rangers, the body parts left by the lions 'would not even have half-filled a plastic shopping bag'. And in August a man was killed at the Punda Maria turn-off south of Parfuri, while ten of his companions escaped by climbing up trees.

It is thought that even more people are killed than are reported officially, for park rangers often find abandoned luggage and torn clothing, but no bodies or even parts of bodies. What the lions leave, the hyenas, jackals, vultures and other scavengers clear up until nothing is left. In fact, incidents are only reported if there are survivors. Thus, one of the four killings in 1998 occurred on 22 July, when an 11-year-old girl was found wandering along the S60 road near Klopper-fontein, south of Pafuri and east of Punda Maria. She had been with her mother and two sisters when they stumbled into danger, and survived by hiding in the hole of a burrowing animal. Whether they were attacked by lions or trampled by elephants and then scavenged by lions and hyenas is not clear. Three days later the mother's body was found with a buttock chewed away and her head distorted, indicating a probable elephant attack, followed by a bout of scavenging.

The problem is that the entire eastern boundary of the park runs along the border with Mozambique, and an estimated 15,000 people have been caught trying to make the trek through the bush and have been repatriated, with more than 4,000 discovered since the beginning of 1997. The actual number of casualties must be far greater (including victims who are attacked by crocodiles as immigrants attempt to cross rivers), and some prides of lions and accompanying scavengers have recognised the easy pickings.

Lions are basically lazy animals, and once an individual kills and eats a person, it seems to prefer what amounts to convenience food that is easy to catch. An entire pride of five lions at Kruger had to be put down as they had taken to man-eating.

The second problem for the park has been that, unlike tourists who must remain in the vehicles except in the secure parts of designated rest areas, park patrols are often out on foot. To date, only one park ranger has been attacked by lions, but he was badly hurt.

Rangers had been attacked previously at Kruger. In the 1930s Harry Wolhunter was attacked twice by lions, on one occasion fighting and killing a male lion with his hunting knife. There also have been man-eating incidents. In 1937 and 1938, for example, several people lost their lives. At Tshokwane and Letaba people were killed and eaten by lions, and staff at the Duba railway siding had a narrow escape when a lion patrolled the area. One man slammed the door of his hut in a lion's face and survived.

Many of these attacks were by old and infirm animals, the so-called grannies of a pride that had worn teeth and were in very poor physical condition. One such female was the man-eater of Pabene. A number of children were playing in the yard when they heard a lion and ran for the safety of their houses. One of them, a 12-year-old girl, stumbled and in an instant the lion was upon her and dragged her into the bush. Her parents and friends took burning straw and scared the lioness away, but when they reached her, she was already dead.

Unfortunately, the local ranger was absent from the area, and so a hunting party from the village went in search of the culprit. They chanced upon a male lion which had a leg injury, and thinking it the man-eater, they shot it. The following day, the dead girl's father was repairing the road leading to the village so that the funeral vehicles could reach his house when he was attacked by the killer lioness. He fought hard and managed to stab the animal in the side of the face and the flank. He received serious gashes to the face and groin for his trouble, but he lived. The lioness slunk away. Bleeding profusely, the man crawled home and was rushed to hospital.

Two park rangers arrived and started to follow the trail of blood from the wounded lioness. They had not gone more than about 60 yards (60m) when they found the remains of another child who had gone missing a few days earlier. Her parents had not reported her disappearance because they thought she had simply left home and had gone to look for work. Rather than track the animal, the rangers decided to put out a bait and wait. They killed a waterbuck and waited for the lioness to arrive. At precisely 8 o'clock that night an emaciated lioness appeared. She had obvious wounds in her shoulder. Two shots rang out, and she was dead. Taken back to Skukuza, the state veterinary officer took a look inside the body, and the grim reality of the animal's bout of man-eating was laid out before him. In the stomach were the soles of human feet, fingers, toenails, bones and fragments of clothing.

She was riddled with parasites. Clearly hunger and an inability to track and catch her normal prey had forced her to take to eating people.

# CHAPTER 6

---

# CATS: SILENT HUNTERS

Lions are not the only dangerous big cats in Kruger. Leopards (*Panthera pardus*) have been active too. Whether a supply of refugees has given them a taste for human flesh or whether they have realised that people on foot are easier game than an impala is not clear, but in August 1998 the reality of man-eating leopards was only too apparent.

A 25-year-old tour guide was taking out a night-time game-viewing group of tourists near Malelane, in the south of the Kruger. He stopped the vehicle on a bridge and had only taken a few steps when the leopard leaped on to his back, causing him to drop his rifle. The tourists tried to scare the animal away but to no avail. They drove to a nearby camp for help, but when the rangers arrived back at the attack site, they found the leopard feeding on the guide's body. The leopard was shot.

In October 1998 one fearless leopard managed to maul six people before being despatched by a pick-up driver wielding a screwdriver. The animal first hijacked two cyclists on their way to work close to the park. It tried to drag one into the bush but let go of its victim and was chased away. The driver of a passing pick-up truck, known locally as a bakkie, pulled up near by. He had seen blood on the road and spotted the two abandoned cycles. He stopped to see if he could help. Immediately, the leopard leapt on to the bonnet, over the cab and into the back of the truck where four terrified passengers were sitting. The driver grabbed a screwdriver, leapt out of his cab and stabbed the leopard to death.

Elsewhere in Africa leopards have left their mark. Late in the 19th century, for example, Romolo Gessi – a lieutenant of General Gordon – had one of his men taken by a leopard from a camp in the Sudan. The body was found the following day hanging half-eaten in a tree. Leopards in Africa have a habit of stashing their kills in trees in order to avoid them being stolen by large scavengers, such as lions, hyenas and hunting dogs.

Lions and leopards were responsible for 15 human deaths in Kigezi, Uganda, in 1929. The attacks coincided with an expansion of the human population in the region.

In the 1940s, Dennis Gibbs, a game warden in south-eastern Nigeria, reported a five-year period during which a leopard was responsible for the deaths of 'many dozens' of women and young children. It had first become a stock-raider before turning to people. It was shot by Gibbs, and on examining the body he discovered the animal had a deformed leg. It had clearly lost part of a paw in a poacher's trap. Unable to stalk and pounce on its more usual prey, it took to livestock and people.

Twelve children and two adults were killed and eaten by two leopards at Kalingombe in Ubena in 1948. In the Njombe area leopards killed seven people and in Chimala another ten during 1951–2.

On 22 January 1961 two people were mauled by a leopard in Fort Hall Road, in the middle of Nairobi. The animal had been lying low in the reeds and bushes of a swamp area close to the road when the two men passed by on their way home. One man was seriously injured, the other unhurt.

The leopard is smaller than the lion, but it is powerfully built and capable of bringing down large prey. It is a particularly cunning hunter. Its habit of using stealth and the surprise attack makes it a potentially dangerous animal to anybody alone in the bush, especially after dark. The leopard generally hunts at night and, from time to time, in some parts of the world, unsuspecting people have been its target.

One observer noted that 'if the leopard were as big as the lion, it would be ten times more dangerous'. Dunbar Brander stated that 'no one is safe' from a leopard, and that 'owing to their knowledge of man's way and being habituated to enter villages at night, they will enter a hut and drag out their victim from his cot'. He went on to conclude: 'From a man-eating leopard . . . people have no security at all.' Events during the past hundred years have borne out Brander's remarks.

Brander was more familiar with the Indian sub-continent than with Africa, and it is here, more than anywhere else in the world, that leopards must be added to the growing list of man-eaters. In fact, in the early days leopard predation on people was often confused with attacks by tigers. This is what happened in Seoni district in central India in 1857, when Robert Sterndale arranged a beat to drive out what he thought was a man-eating tiger from the bush. It had killed a local man. As the drive was in full swing a leopard appeared but the hunters ignored it. Returning to camp after the drive, Sterndale examined the remains of the victim more closely and realised his mistake. The culprit was a leopard, probably the animal which was flushed out during the drive. His error cost over 200 people their lives, for the leopard – the 'Kahani Man-eater' – terrorised an 18 mile (29km) diameter patch of countryside for the next three years. It escaped several organised hunts with such ease that the superstitious local population considered it to be a were-leopard. It was killed eventually by accident by a local hunter who mistook it for a pig.

One of the more infamous leopards was the 'Man-eater of Panar', which took more than 400 people before it was shot by Jim Corbett in 1910. But, some years later – between 1918 and 1926 – it was another remarkable leopard that hit the world's newspaper headlines, and even became the concern of the British Parliament. It was the man-eating leopard of Rudraprayag.

The leopard shared its territory with about 50,000 people, and crossing the area was a pilgrim route through the Garwhal Hills. Thousands of people travelled it on foot annually to reach Hindu shrines in the Himalayas. The animal began its man-eating habits outside a small village just 12 miles (19km) from Rudraprayag, at the place where two rivers merge to become the Ganges. It was just after the 1918 influenza pandemic, when a million people died throughout India. Cremations on the banks of the river in this mainly Hindu region were so numerous that the local community was overwhelmed. Many bodies were incompletely burned and the leopard, being an opportunistic scavenger as well as an accomplished hunter, took advantage of the windfall. It acquired a taste for human flesh. When the disease receded and things began to get back to normal, the supply of corpses dried up and the leopard took to killing live people.

Throughout the nine-year period, the local people underwent a self-imposed curfew each night, locking and bolting doors and windows. Such was the terror, nobody dared step out. But even those locked away were not safe. Jim Corbett, in his book *Man-eating Leopard of Rudraprayag*, describes how a 14-year-old orphan boy and his flock of 40 goats were locked away in a small room each night by the flock's owner, the boy's master. There were no windows in the room, and the single door was 'locked' with a slither of wood jammed through a catch. The boy placed a stone against his side of the door. One night, the leopard came. It clawed at the door, dislodging the wood, and then pushed hard against the rock, shoving the door ajar. Once inside it ignored the goats, but took the boy as he slept. The goats escaped, but none was harmed. In the morning the goats' owner found the boy's remains in a deep ravine close to the village. Despite living and sleeping immediately above the room, the man heard nothing.

This ability of the leopard to hunt and kill silently was illustrated by a tragedy from another village. This time two men sat in a dark room smoking from a hookah (an oriental tobacco pipe in which the smoke is drawn up by a long tube through water). The door to the room was closed but not locked. One of the men dropped the hookah, spilling charcoal and tobacco on to the rug on which they were sitting. The other man bent down to clear the mess, and just as he looked up he saw that the door was open. In the doorway, silhouetted by the moonlight, was a leopard. It was carrying the other man, and he was already dead. The survivor had heard nothing, even though they had been sitting but an arm's length from each other.

A third story echoes the theme. This time the wife of a village headman was sick with fever, and two of the woman from the village were asked to stay with her during the night. They slept alongside each other, the sick woman in the middle. At midnight or thereabouts, the leopard entered a small window, avoided a large brass vessel containing water on the window sill, and killed the invalid. It tried to pull the body out the same way, knocking over the water container in the process. The rest of the household woke up immediately, but the leopard made its escape. The woman's dead body was found under the window. On her throat were four deep tooth marks.

The leopard's cunning is illustrated by a fourth tale. Two brothers and the 12-year-old daughter of one of the men were moving a herd of 30 buffalo from one grazing site to another. As night fell, they made camp and, after a meal, they put their blankets down between the tethered herd and the road, and went to sleep. They were woken by agitated buffalo in the early hours of the morning. The two men lit a lantern and went among the herd to calm them. They had only been away for a few minutes, but in that time the leopard had slipped into the camp and taken the girl. All that was left on the blanket was spatters of blood.

On another occasion, several pilgrims arrived at a village and, as it was close to dusk, they asked if they could sleep on the veranda of the local shop. The shopkeeper advised against it, warning that the man-eater frequently passed this way. He suggested they travel the four miles further on to the next pilgrim rest site. But the pilgrims were tired, and wanted to eat and rest where they were. The shopkeeper was adamant, and at first refused them permission. Just then, a sadhu or holy man arrived, and said he would sleep alongside the pilgrims. If the leopard came in the night, he bragged, he would tear it limb from limb. The shopkeeper relented, and the pilgrims, with the sadhu in their midst, settled down for the night. In the morning, the sadhu had gone, his bedding splashed with blood. The shopkeeper knew what had happened, and together with the pilgrims followed a trail of blood across three terraced fields to a low stone wall. Laying across the wall was the sadhu's body, the lower part eaten away.

The first of these horrific attacks took place at Bainji village. The date was 9 June 1918. By 14 April 1926, at Bhainwara, where it made its last kill, the animal had accounted for at least 125 pilgrims and villagers. The leopard itself is said to have had nine lives, for despite a price on its head it survived many attempts to trap and kill it. Hunters and soldiers in the area were rallied to track and shoot it. Sportsmen from all over India were encouraged to join the hunt. Traps, baited with dead goats, were erected at key sites along the pilgrim road, but it evaded them all. At various times it avoided food contaminated with different types of poison, dodged shots from guns rigged to trip wires placed around carcasses, escaped from a box trap by digging its way out, and side-stepped a fusillade of bullets from two British army officers who ambushed it crossing a suspension bridge. On one occasion, it was caught in a steel spring trap, but its leg was caught at just the point where some of the teeth had corroded and, after a struggle, it was able to pull itself free before hunters turned up to shoot it. On another, it was sealed up in a cave. Watched by 500 gawking faces, a local VIP opened the entrance whereupon the leopard burst out and charged into the crowd.

One particularly nasty method employed to catch man-eating leopards was the insertion of a small bomb into the body of an animal killed by the predator. When the leopard returned to the carcass to feed, its teeth made contact with the bomb and it exploded. Either its jaws were blown off and the animal killed instantly, or more than often it crawled away to die a slow and painful death.

The Rudraprayag leopard avoided the bombs, guns, traps

and poisons. It gained such notoriety that questions were asked in the British Parliament, and Jim Corbett was invited by the Deputy Commissioner to hunt down and kill the beast. The chase was to last several months, for the man-eater operated in a 500 square mile (1,295 sq km) area of rugged mountains. The area was dissected by a fast-flowing river, the Alaknanda, and the two banks joined by two suspension bridges and a rickety rope bridge. Corbett reasoned that if he prevented the leopard from crossing the bridges, by blocking the entrances during the night with thorn bushes, he could confine it to one side of the river and half the area he had to cover. This he did, but not before the animal attacked a woman in her home. She was collecting together the dinner pots and plates after the evening meal, when her husband heard a clatter. The leopard dragged the woman down a lane between several houses. On hearing the commotion, the neighbours instantly slammed their doors shut and locked them. Meanwhile, the leopard killed its victim and carried her to a nearby ravine. The villagers discovered her remains in the morning.

At one point, Corbett killed a male leopard, but was not convinced it was the man-eater. Events were to prove him right. While the villagers on one side of the river celebrated the death of the leopard, a young 18-year-old girl on the other side, who had stepped outside her house to relieve herself, was killed and partially eaten. One of the suspension bridges had not been blocked that night and the leopard had crossed over. The man-eater was still on the loose. Corbett and his companion tracked the leopard, but also were stalked by the animal itself. They risked life and limb in their attempts to eliminate it. In the meantime, the killings and maulings continued. One victim had a lucky escape. A woman and her baby were asleep in a small room, when the leopard broke in and grabbed her arm, trying to drag her out of the room. Fortunately, she did not faint and had her wits about her. As it backed out of the room, dragging her across the floor, she slammed the door on it. She escaped with a lacerated arm and deep wounds on her chest.

In a neighbouring village a man sitting on his veranda was

not so fortunate. The leopard grabbed him by the throat and dragged him 400–500 yards (400–500m) to a hollow surrounded by brushwood. Before it was disturbed, the man-eater had eaten the throat, part of the jaw, shoulder and thigh. Corbett laced the remains with cyanide, in the hope that the leopard would come back the following night. Inspection of the corpse in the morning revealed that the man-eater had, indeed, returned. It ate the other shoulder and leg, but avoided the places where the poison had been placed. The following night, more cyanide was placed on the remains, and Corbett staked out the site, sitting all night in a hide with his rifle at hand. Sure enough, the leopard returned but succeeded in eating its fill, bypassing Corbett, and visiting the river to drink. The leopard had eaten meat, bones, poison and all, but it was still alive, and it was ready to kill again.

A 70-year-old woman stepped back inside her house and was about to close the door, when the leopard sprang and dragged her from the doorway. Still alive and screaming, she was hauled through the village, but not one villager came to her assistance. They were too frightened. When she was dead, the leopard carried her – rather than dragged her – across waste land, an open ravine, up a hill for a hundred yards (90m) or so, and on to some flat, open ground. Corbett found her body, flecked with white flower petals.

The leopard, by this time, had become very bold indeed. Failing to catch people outside their homes or through open doorways or windows, it took to scratching against doors and actually breaking in to houses. This happened to a woman sleeping alone in her house. The leopard grabbed her by the left leg and began to drag her to the door. With presence of mind, the woman grabbed a tool for chopping cattle chaff and hit the attacker. The animal failed to release her, and as it backed out of the door, the door slammed. But it still had her leg, and the leopard was on one side of the door while she was on the other. Such was its strength, it gave a great tug and the woman's limb was torn off. Its days, however, were numbered.

The leopard's last human victim was at Bhainswara. A man, his wife and several children were bringing water back from the stream for the evening meal and were climbing the steps to the house when the couple's son, who was bringing up the rear, dropped his brass vessel. The mother turned to chastise him, but he was not there. Then the father saw the blood marks on the ground and followed them to a low wall and across a field. The body was discovered the next morning. When Corbett arrived, he found that the boy's neck had been dislocated, which meant that he had been killed instantly. Noise from the village had prevented the leopard from eating any of its victim, but Corbett thought it would return to finish its meal so he used the body to lure the leopard into the village. He would be hiding near by with his rifle. The corpse was lashed by a chain to a stake buried in the ground. Corbett lay in wait on a veranda. After a fright from a small kitten which had crawled over his prone body, he heard the unmistakable commotion of two leopards fighting. The man-eater had returned to the kill site, but had been intercepted by the resident male. Corbett's vigil had been in vain.

Heading for Golobrai, Corbett decided he would make a last-ditch stand beside the road between Rudraprayag and Golobrai. The pattern of attacks on people and livestock made him realise that the leopard must travel that way at least once every five days. He put up a shooting platform in a mango tree not far from a pilgrim rest house, tied a live goat with a bell to a stake in the middle of the road, and sat down to wait. Ten nights he waited, and then on the eleventh, the leopard appeared. He heard a rush, and there in the rifle sights were the leopard's shoulders. He fired, listened and heard nothing. It was 10 p.m. Corbett then had to wait all night before he could get down and check whether he had made a hit. In the morning, as the light came up, he saw that the goat was still alive, and there was a streak of blood on a low rock. Following the trail of blood, Corbett found the dead leopard with its rear part in a hole.

Corbett had killed the leopard of Rudraprayag. It was a particularly large, old male, way past his prime but nevertheless unusually strong. Its length between pegs was 7ft 6in (2.3m). The goat used to lure it to its death became a celebrity. It was given a brass collar and became a source of income for the proud owner. A plaque was placed where the leopard had fallen. The date was 2 May 1926, and even today folk memory stretches back to Corbett Sahib who saved the local community from the killer.

During his tracking of man-eating predators, Corbett often had to contend with the deeply rooted superstitions of local people. Just as Africa had its were-lions, India had its were-leopards. Indeed, while Corbett was tracking the man-eater of Rudraprayag, he was involved in an incident with a supposed suspect.

In Garwhal, attacks on humans by man-eating predators are attributed to sadhus who are believed, as Corbett put it, 'to kill for the lust of human flesh and blood'. In some villages it was thought that the leopard of Rudraprayag was a were-leopard and it almost cost a sadhu his life. His village had been terrorised by the leopard and the villagers were under-standably both frightened and angry. They took out their wrath on the sadhu and were about to kill him when the Deputy Commissioner, who was camped nearby, chanced upon the lynching. Stepping in to pacify the frenzied crowd, the Commissioner proposed that the sadhu be placed in prison and his guilt established properly. The villagers agreed and he was guarded day-and-night for seven days. On the eighth day, word came from a neighbouring village that a house had been raided and the occupant dragged away by the leopard. The villagers agreed to release the sadhu. They had caught the wrong culprit.

Superstitions such as these are still apparent in rural areas to this day, and the leopard attacks have not stopped. More recently, Garwhal became the focus of world attention once again. From 1993 until 1998, women and children were the target for leopards operating in Pauri, one of Garwhal's four districts. Currently, attacks occur once every ten days on average, and the number of deaths from leopard attacks has been increasing year on year. In 1993, the district had five on the official records, but this had reached 17 per year during 1996, and in 1997 there were 19.

The problem is the same as that faced by the tiger. Forests, the leopard's natural living space, are being cut down by illegal logging operations and exploited by corrupt timber merchants. Prey is scarce. People, and children in particular, are a viable alternative. Attacks follow a similar pattern. A woman is walking home in the dark with her youngsters alongside. The leopard jumps out of the bush and knocks one of the children to the ground, grabs it and runs off. According to newspaper reports, this is precisely what happened to a young villager on her way home with a baby in her arms and her eight-year-old daughter following behind. The leopard jumped on the young girl and, snarling viciously, tried to pull her away. The mother caught the girl's arm, but in the tussle slipped and fell down an embankment. She lost her grip and the girl was carried off. Villagers heard the screams and came with flaming torches made from rags drenched in kerosene. The girl's half-eaten body was found under bushes a short walk from the village – another victim to add to Pauri's escalating statistics.

It is thought that 187 leopards are living in the Garwhal district and since 1993 about 20 have been declared man-eaters. They go for the thinnest and weakest members of the community. In September 1997, a 24-year-old women stepped out of her house to go to the toilet and was killed and dragged off by a leopard. Her husband first found her shoe, which led to her half-eaten body. Others have been luckier, like Siddheshwari Singh who was caught by the head and dragged several yards. She managed to grab some dried cow dung and hit the leopard and it let her go. She still has the scars from its initial bite.

Elsewhere in the leopard's range, the story is repeated. The human inhabitants of Nepal, for example, have to contend not only with tigers, but also leopards. Nepalese children are clearly most vulnerable, and at the end of 1986 and beginning of 1987, ten children were killed by tigers near one village in the Pokhara district of western Nepal. On one occasion, a tiger snatched a child from its mother's grasp and made off with it into the forest. The animal was eventually tracked down and shot by Royal Palace hunters. Adults, though, are not immune. Early in 1989, a leopard mauled several people and actually killed a woman at Kaski, in the central hills of Nepal.

Later in the year, a leopard killed a 70-year-old man who had been collecting firewood from the forest near Pokhara. Local people tracked it down and stoned it to death.

Other parts of India, too, have leopard problems. In October 1972, a leopard killed three boys aged four, seven and twelve within the space of eight hours in villages near Junagadh, in western India. India's Bihar state has also seen leopard attacks. In May 1982 a leopard roaming the forest district of Hazaribagh claimed its 41st victim – a girl aged six.

In southern Bangladesh, leopards as well as tigers are active predators, preying on people. In April 1986, a rogue leopard killed three children. Hunters tried to track and kill it, but only wounded it. The enraged animal then went on to kill a woman and badly mauled 12 men.

In Sri Lanka, the island's own sub-species of leopard has indulged in man-eating on two occasions. In 1924, there was the man-eater of Punani. It was an enormous beast which was eventually shot by Shelton Agar. Its stuffed and mounted body can be seen in the National Museum in Colombo. And, in the 1950s, the man-eater of Pottana patrolled the pilgrim route along Potuvil Kataragama, but little is known of what mayhem it might have caused.

That people are targeted by leopards should come as no surprise, for they have been attacking our near relatives for thousands of years. In certain parts of Africa leopards often take gorillas. In the Virunga mountains, on the borders of Rwanda and Zaire, the very rare mountain gorilla is subject to leopard attacks. American biologist Dale Zimmermann actually watched a black (melanistic) leopard track a group of gorillas at an altitude of 12,000ft (3,600m) on Mount Muhavura. Not long before, a tracker had found two dead gorillas in the same area. They had been mauled by a leopard.

There are other black, leopard-like predators, reported in Africa, but these fall into the category of crypto-beasts rather than proven reality. One is the *ndalawo* of the Uganda forests. It has a similar shape to the leopard but has black fur on the back and grey on the belly. There is also the *nunda* or *mngwa* (meaning 'strange one') which is slightly larger and lives only in dense forests along the coast of Tanzania. Like the Nandi bear, both these strange cats have evil reputations as man-eaters, and they are much feared by the local people.

The *mngwa* is said to be the size of a lion and striped grey, and although thought to have the same status in African mythology as the dragon in Europe in the Middle Ages, there was a time when it moved from children's fairy tale to government report. In 1922, local magistrate Captain William Hitchens came across an unusual incident. In the Lindi market, traders left their goods in the market place overnight where they were guarded by several local policeman who worked in shifts throughout the night. One man went to relieve his colleague at midnight, only to find him dead and badly mutilated under one of the market stalls. The policemen thought their companion had been attacked by a lion, but lions rarely came into the town. Clasped in the dead man's hand, however, was a clump of grey fur. Sometime later, two people said they witnessed the attack. They described a big cat, the size of a donkey, and coloured like a domestic tabby. There was no doubt in everybody's mind that they had been visited by a *mngwa*. Hitchens and the local police chief admonished the other police officers for believing in such superstitions, but the dressing-down was premature. That night, another constable was attacked. In his hands and around the buckles of his uniform was matted grey fur.

For several weeks, the beast continued to attack people in the fishing villages along the coast. Pit traps and poisons were used in an attempt to kill it, but it was never seen. Then, the killings stopped as suddenly as they had begun, and no more was heard of the *mngwa* until 1937. One day at Mchinga, a small fishing village, a local hunter was brought in from the bush on a stretcher. He had been badly mauled by a big cat that he firmly identified as a *mngwa*. Bearing in mind that a hunter is likely to know the difference between a lion, leopard, hyena and something altogether different, Hitchens was inclined to believe him.

European hunter Patrick Bowen was also convinced that he had once been on the trail of a *mngwa*. He told Frank Lane, who was writing in *Nature Parade*, about an animal that had carried off a small boy from a village. Bowen and another hunter went in pursuit, following the animal's spoor. At first they thought they were tracking a lion, but when they crossed some hard wet sand they realised that the clear pug-marks were not made by a lion. They resembled that of a leopard, but were made by a leopard the size of a lion!

A stockier version of the leopard that is firmly in the realm of reality is the jaguar or el tigre (*Panthera onca*). It lives in South and Central America where it grows to 6.2ft (1.9m) long, weighs up to 250lb (113kg), and has the strongest jaws of all the big cats. While lions and tigers tend to kill by strangulation, the jaguar takes the head of large prey into its mouth and, with the large canine teeth in the upper and lower jaws brought together, punctures the cranium and pierces the brain. It can penetrate bone up to a half-inch (1.3cm) thick.

Fortunately, in recent times, it has rarely been implicated in deliberate man-eating attacks on people, although there have been a few stories from the past. Travellers, such as J. W. B. Whetham in Central America and Leonard Clark in Yucatan, have heard of reports of jaguars attacking people from behind and killing them by biting into the skull or severing the backbone. In some areas, native people were in dread of them.

W. R. Stevenson, who lived in South America for many years in the mid-19th century, writes in his *Historical and Descriptive Narrative of 20 Years Residence in South America* about jaguars staking out villages at night and waiting for people to leave their houses. He reported that at Guayaquil hungry jaguars had become man-eaters. Similarly, reports from the Nahuel Huapi National Park in western Argentina tell of jaguars circling villages at night in the hope of ambushing a person foolish enough to leave their house.

The Boteudos on the Averanha considered jaguars to be man-eaters, and the local people on the Tuarutu and Vindana ranges of what was British Guiana told explorer R. Schomburgk that jaguars would attack people just as readily as they killed cattle and tapirs. Schomburgk also writes how jaguars entered villages and took children, and that the Arekuna Indians in Guyana built palisades around their huts to keep jaguars out.

In some parts of Peru, according to J. von Tschudi, local

people had moved their villages on account of attacks by jaguars. He heard of one village – Mayumarea, on the road to Anco – that was abandoned for a hundred years because of predation on humans by jaguars.

A. R. Wallace, travelling in the Amazon in the late 19th century, reported hearing the story of a jaguar entering a hut and the occupant being attacked in his hammock. And, W. N. Herdon tells of a Quichua Indian living on the banks of the Espirtu Santo river near Rio de Janeiro having to live in his loft because a jaguar kept entering his house at night. He was sure it would have killed him on sight.

The most peculiar story to come from these early European travellers, however, must surely be that from Lt H. L. Maw. He tells how a black (melanistic) variety of jaguar regularly entered the plaza of a Peruvian pueblo at midday and seized the first person it could catch. It attacked and killed 50 people, dragging them into the surrounding countryside to be eaten, before it was killed itself.

F. de Azara also reports that the few jaguars which remained in Paraguay while he was there in the early 1800s killed and ate six people. Two people were carried off while warming themselves by a campfire, even though they were sitting alongside several able-bodied companions.

Charles Darwin also reports in his *A Naturalist's Voyage round the World in HMS Beagle* that woodcutters were killed by jaguars on the Rio Parana and Rio Paraguay. And many years later, Marquis de Wavrin mentioned that jaguars entered huts along the Rio Putumayo and carried off human victims. All agreed that although jaguars avoided contact with man in the forest, they had become increasingly bold in places where they had got used to heavy river traffic. They even entered ships, and there is the tale of a man who was coming up from below being seized when he arrived on deck. He escaped death, but lost his arm in the attack.

He also wrote about the way jaguars were dangerous at the time of floods. They were driven from islands in the river and could easily land up in a settlement. Just such an event took place on 10 April 1825 at the convent of San Fernando. Explorer J. R. Rengger arrived there a few days after the incident and recorded it in his book *Naturgeschichte der Saeugethiere von Paraguay*. The convent was situated about 18 miles (29km) from Sante Fe on the banks of the Rio Grande. Periodically, the river flooded and the small islands close to the convent were covered by water. Wildlife, including a jaguar on this occasion, swam to the nearest land – that on which stood the convent. Imagine the surprise of one of the lay brothers when, having completed his confession and prayers, he entered the sacristy and on opening the door came face to face with the jaguar. The animal grabbed the man and dragged him to a corner of the room. In Rengger's account, this was the only casualty, but the story has been embellished over time. Later versions tell how the caretaker, hearing the commotion ran into the sacristy and the jaguar killed his second victim. More people entered the room, and another was killed. A senator – Senator Iriondo – staying at the convent tried to enter from a room on the opposite side, but he chose the wrong moment. Just as he went in, he met the jaguar coming out. The crowd which had now gathered heard him cry 'Here it is!', followed by 'Here he goes!' and then the immortal words 'Save me, save me!' drowned out by the jaguar's snarls. The jaguar had claimed its fourth victim. Eventually, it was shot.

An entertaining postscript to the story occurred some years later when a researcher decided to investigate the story. He visited Santa Fe, only to find that there was no convent and nobody had heard of the characters who had been involved in the jaguar incident. Writing up the work, the researcher concluded that the entire story had been a fabrication. Darwin and other eminent scientists had been duped. Unfortunately, the young author failed to realise he had visited Santa Fe, New Mexico, in search of the facts and not Santa Fe, Argentina!

A quarter of a century later in 1850, another incident illustrated the ferocity of the jaguar. The story is told by Richard Spruce in his *Notes of a Botanist on the Amazon and Andes*. He tells how two men were walking though the forest when a jaguar leaped on to one of them. The man was well-built and wrestled with the beast, but it had one paw free and it scalped him. A third man arrived and killed the cat. The victim wore a skull cap for many years afterwards because the wound was so sore and slow to heal.

On a few occasions, it is thought that female jaguars with cubs have attacked people while defending their litters. Having sampled human flesh and realised how easy people are to kill, the animals had gone on a man-eating spree. E. B. Prado recalls a jaguar mother behaving in this way in his *The Lure of the Amazon*. The events took place in the lower Aripiana region, and Prado was asked to hunt down the man-eater, a female with an injured paw. In the course of 20 days, the jaguar had killed six people. One victim was a woman washing her clothes at the river's edge, in the company of her pet dogs. The jaguar struck, killing the woman and two of the dogs. Two woodcutters heard the screams and barking dogs; they came running and the jaguar ran away. The men went to find help, but when they returned the body had gone.

Prado tracked the jaguar to her den beside a waterfall and pool where deer came to drink. He hid in the cactus scrub all night and watched. For a few moments he saw the mother with her cubs, and then nothing. As the sun came up and the heat of the day made Prado drowsy, he suddenly became alert as some pebbles fell at his feet. Looking up, he saw the jaguar and her cubs. The cubs were playing and had loosened the rocks. The mother spotted Prado's movement and coiled up ready to spring on him. She leaped, Prado grabbed his rifle and fired, and the man-eater was dead. Examining her body, he found that one paw was deeply scarred.

US President Teddy Roosevelt, writing in *Through the Brazilian Wilderness* at the turn of the century, mentions the testament of an Argentine captain whose camp was raided by a jaguar several times. The cat was after the dried beef stored there. When the meat was put out of harm's way, the jaguar changed its tactics. Moving silently through the camp, without disturbing the dogs, it seized one of the men, killing him by biting through his skull. In the ensuing commotion, it dropped the man and made good its escape. The dogs tracked it the following morning and it was shot. Examination of the body revealed it to be a large male in good condition. He also tells of another male in prime condition that had taken to killing people in the Chaco. It killed three people, eating two.

Natural historian and traveller Leo Miller also discovered a jaguar that had taken to man-eating at a rubber camp. The animal had stalked a two-year-old child, but was spotted by the mother who despatched the jaguar with a machete.

Leonard Clark, who wrote *The Rivers Ran East*, tells of a small female jaguar, weighing an estimated 150lb (68kg), that pulled a man from an ox-cart and dragged him into the bushes. The two were found later. The man had clearly fought for his life. The jaguar was strangled, and the man was unconscious and close to dying. Clark also mentions a three-legged jaguar that took to man-eating in the Opaxta district of the Yucatan. It followed a regular circuit around the district, completing it every seven or eight days. During that time it killed five men. He also recollects walking through the cemetery at La Merced and counting 32 graves of men who succumbed to attacks by jaguars. Whether these were all genuine attacks, or whether it was a way in which to settle old scores without persecution, we shall never know.

In general, people living in jaguar country were in awe of the animal. It featured in religion, myth and legend. In some societies it was regarded as an ancestor and in others as a malevolent spirit with which it was useless to pick a fight. Thus, man-eaters were sometimes treated with a curious indifference. The Putumayo Indians, at the turn of the century, for example, were plagued by an unusually fearless population of jaguars. The cats would creep into the large communal houses in which a hundred or more Indians had congregated and select a victim. As the jaguar dragged the person away, the rest of the group did not lift a finger to help. Instead, they called out to the beast that it should take the child or woman and not come back. The jaguar departed unmolested with its victim.

Throughout South America the jaguar was considered a god, but the worst excess of cat worship must have been exhibited by the secret society of the Nahualists from Mexico. Like the lion- and leopard-men of Africa, these people took on the jaguar's shape, wore its skins and carried out ritual killings, often eating part of their victim.

Fortunately, real jaguars appear not to be persistent man-eaters. There may be instances when a child or a person on foot and alone in the forest is taken more by mistake than by design; otherwise jaguars and people tend to go their separate ways.

The same can be said of the elusive mountain lion, cougar or puma (*Felis concolor*). The mountain lion is a member of the cat family that lives in the more rocky and mountainous areas throughout the New World, from Canada in the north to Patagonia in the south. Though more slender than the jaguar, the mountain lion is nonetheless a powerful predator more than capable of bringing down and killing a person. It has a head and body length of up to 6.6ft (2m) and weighs over 220lb (100kg). It is active at twilight, and humans have been on its menu since mountain lions and people began to share the same country.

North American Indian stories identify the mountain lion as a wild beast that is dangerous to man, and there are records of early settlers being attacked as far back as the 1750s. Lee Fitzhugh and Paul Gorenzel, from University of California at Davis, found just such a case (among 66 others between 1750 and 1985 which they included in their report) in Pennsylvania. The year was 1751, and the only record that remains of the event today is a tombstone with a carving of a mountain lion above the name of the unfortunate victim.

One of the first recorded attacks, however, was on the famous Lewis and Clark expedition across North America. In 1805, one of the team was attacked by what was described as a 'tyger cat', but just as it sprang it received a shot from a carbine for its trouble and slunk away.

At the turn of the century, Teddy Roosevelt wrote in his *Outdoor Pastimes of an American Hunter*, published in 1905, about several known attacks on Indians, cowboys and, in one instance, a three-year-old child. There were many others.

Don Zaidle – hunter, fisherman and wildlife author – uncovered several old references from his home state of Texas. In 1850, for example, a woman was attacked in her bedroom. She was checking out why a hen brooding eggs under the bed was making such a noise and came face to face with a mountain lion. She turned to run but the lion leaped and grabbed the back of her neck. Two men in the next room came to the rescue and shot the cat dead before it could do more damage.

Some twenty years later a school teacher in southern Texas was attacked after having dinner with neighbours, and in 1908 another woman was attacked actually in her bed. She had been recovering from an accident. As a consequence of these and many other incidents, such as the cavalry soldier who was wounded in the Apache Wars but was subsequently killed and eaten by a mountain lion while trying to leave a water-hole, the mountain lion was considered vermin and fair game for the hunter's gun. Many were killed.

After a spell in the 1950s when the mountain lion was thought to be on the brink of extinction throughout North America, it seems to have gained a new lease of life and its numbers now appear to be rising. In California, Colorado and Idaho, for example, where the largest population of mountain lions is thought to reside, it is estimated that there are up to 12,000 individuals living today. And mountain lions have been seen in states from where they were thought to have disappeared, such as Maine.

Although the North American mountain lion population is on the increase, unfortunately its living space is not, so in some towns and cities, particularly those in California and Texas, mountain lions and people are coming into conflict. They are often seen in hilly residential areas throughout the San Fernando Valley. Residents of Reyes Drive, for example, are frequent observers of mountain lions and coyotes. And, in Canada's British Columbia, mountain lions have been turning up in the suburbs of Vancouver and the town centre of Victoria. There, in September 1998, a mountain lion made it past the receptionist at Scott Plastics and into the building where an employee managed to trap it in an office. It was tranquillised and removed, and was found to be grossly undernourished.

Mountain lions have also been seen in the vicinity of the town suburbs at Port Arthur and Tyler in Texas, a shopping mall at Montclair, California, the town centre at Fort Worth,

Texas, the hospital at Placerville, California, the 14th hole of a golf course at Chico, California, numerous California schools, and a whole lot more. Schools in El Dorado County and Cameron Park, and the campuses of California State University at Sacramento, University of California at Santa Cruz, and University of California at Davis have all had mountain lion alerts. One student at Santa Cruz was ambushed by a mountain lion when jogging, but escaped by jumping over a fence and stopping a passing motorist.

The problem is that mountain lions are territorial animals, and on average require about 100 sq miles (259 sq km) of space each. The male has the larger patch, overlapping with the territories of several females with which it will probably mate. Although a female's territory might overlap with that of other females, young males are intolerant of each other and need to eke out their own places. With a premium on space, even on the vast North American continent, animals are inevitably going to move into the suburbs, and it is significant that many confrontations with and attacks on people (about 60 per cent according to Don Zaidle, and included in his book *American Man-Killers*) involve young mountain lions. Also, populations of the mountain lion's normal prey animals are going to be significantly lower in the suburbs than in the wilderness, and so humans are going to be increasingly on the mountain lion's list of desirable foods, there being so many of them available.

It is out in the wilderness, however, that mountain lions are more usually seen. And even here, often as not, the mountain lions come off worse. In one radio-tracking study in southern California of 32 mountain lions fitted with collars, 25 were poisoned, shot or killed in traffic accidents. The Florida panther, a rare subspecies, was in so many road accidents that it was close to disappearing altogether.

Occasionally, however, the human residents and visitors are at risk, and since the 1970s attacks have increased in frequency, particularly in the western half of the continent. Nevertheless, no more than 57 attacks on people, ten of them fatal, were recorded officially during the hundred years between 1890 and 1990 according to a report published in 1991 by Paul Beier, of the University of California at Berkeley.

Unofficially, according to research carried out by Don Zaidle, the statistics should be revised. He suggests there have been far more attacks and fatalities than the records show.

Families walking in the woods have been key targets, the mountain lions taking an interest in the younger and more vulnerable members in a group of hikers. On 2 August 1984, for example, an eight-year-old boy and his four-year-old brother were strolling ahead of their parents in the Big Bend National Park in Texas when they walked straight into a mountain lion. The animal focused its attention on the younger boy, but his brave older brother placed himself in front of him to protect him. They both ran, but the cat leapt and bit down on the older boy's skull, pinning him to the ground. The parents caught up, and the husband grabbed the mountain lion by the neck, wrenching him away. The two struggled, the mountain lion clawing at his adversary, but the man would not let go. Eventually, he kicked at the cat, lifting it high into the air. The two faced each other, the cat spitting and snarling, the man screaming. The cat ran into the undergrowth. The boy was still alive, and the family carried him back to their rented cabin and got him to hospital. The mountain lion tracked them all the way back to the hut.

In 1985, Cuyamaca Park in California was where a family were tracked and threatened by two young mountain lions; and in 1987, two children were attacked by a mountain lion at Ronald W. Caspars Wilderness Park, Orange County, California. Nearby, in 1988, a woman who had visited the toilet in O'Neill Regional Park, also in Orange County, climbed the partition wall between cubicles and hung from a ceiling pipe when a mountain lion pushed its nose under the door. The animal sniffed about but that was all; she came down from her refuge shaken but unharmed. Another woman, aged 51, walking in Butte County, California, was not so fortunate. She disappeared mysteriously in early November, and her mutilated body was not discovered until later in the month. It was not until six years later, however, that the woman's injuries were put down to a mountain lion attack.

The following year, in March 1989, a five-year-old girl was

grabbed by the head by a mountain lion, again in Caspars. She was dragged to the bushes, but a passing hiker helped her mother beat the animal with tree branches and drive it off. The girl survived but her face was disfigured, she lost an eye and was left with permanent partial paralysis. The authorities decided to ban children from the park.

In September 1989, attention switched to Missoula, Montana, where two young mountain lions attacked and devoured a five-year-old boy who had been riding his tricycle not more than a few yards from his front door.

In March 1992, three young brothers and their parents were hiking in Gaviota State Park, California. The boys were about 30 yards (30m) ahead of their parents when a mountain lion seized the nine-year-old by the head and started to drag him into the forest. The father grabbed a stone and hit the cat between the eyes. It dropped the boy who survived after being patched up at hospital. A similar event occurred at Canyon Lake, near Phoenix, Arizona, where a mountain lion attacked a five-year-old child and his father was able to fend it off using stones. Again, the child survived after extensive surgery.

In May 1994, the parents of a three-year-old boy saw off a mountain lion that had charged their child in Cuyamaca Park, California.

In January 1998, just two weeks after the ban on children was lifted at Caspars, a mountain lion that was snarling and hissing circled closely a group of women and children. About 75 people were evacuated and the lion was shot dead.

In June 1998, a woman and her three young daughters were threatened by a mountain lion for about 15 minutes while hiking in the Chisos Mountains in west Texas. She was armed with just a pocket knife, but she threw rocks at the cat and it eventually retreated.

Solitary hikers, hunters and horseback riders are understandably another group vulnerable to mountain lion attack. Mountain lions are powerful animals and a fully grown cat can easily bring down even an adult human. In June 1990, for example, a 28-year-old woman was hiking in Fourmile Canyon, near Boulder, Colorado, when she was chased by two young mountain lions. She climbed a tree but the two cats just came after her. She kicked one out of the tree and jabbed the other in the eye, but the cats circled her tree for more than half-an-hour before they left and she was able to come down and make her escape.

In September 1994, a 15-year-old boy was riding his horse near his house at Shasta Lake, California, when the animal shied. It had smelt the mountain lion and would not budge. The horse bucked and the boy went flying. He had a soft landing, though, right on top of the mountain lion that had been lying in ambush under a bush!

Mountain lions also take a strong interest in joggers. Whether lone people running through the forest present the cat with a target not unlike their natural prey, say, a fleeing deer, is not at all clear. Whatever the reason, joggers are vulnerable. In January 1991, an 18-year-old student from Clear Creek County High School near Idaho Springs, Colorado, went jogging at midday and never returned. His partially eaten body was discovered a couple of days later. A large mountain lion was standing over it. In April 1994, a woman jogger in a California recreational area – Auburn Lakes Trail, to the north of Sacramento – was killed by a mountain lion. The attack occurred on the morning of 24 April when she was practising for a forthcoming marathon. The mountain lion had leaped on her from a bank and the impact had propelled both her and the cat down a steep embankment. The mountain lion's canine teeth pierced her brain and she was killed. The cat ate its fill and covered the body. It was only found because the woman's sun visor was nearby. Similarly, at the Cuyamaca Rancho State Park, southern California, a 58-year-old woman was attacked and eaten while jogging through the forest in December 1994. A 39-year-old man jogging in a forested area in Kings County, Washington, was luckier. A mountain lion shot out from behind a tree and gave chase, but the man spotted it, yelled at it and threw stones. The mountain lion backed off. The date was January 1996.

Of the smaller species of wild cats, only the fishing cat (*Prionailurus viverrinus*) of southern Asia has been implicated in man-eating. It has a head and body length of no more that 34in (86cm) long, stands 16in (40cm) at the shoulder, and weighs a maximum of 31lb (14kg). As its name suggests, it catches fish, as well as making a meal of freshwater invertebrates, such as ampullaria snails, frogs, snakes, small mammals and birds. It has also taken to baby-snatching, that is, if a missionary on the Malibar coast is to be believed. He claimed that the fishing cat entered huts and dragged away small children. The observation was greeted with disbelief, but a look through the historical record reveals that Robert Sterndale also wrote about an incident with a wild fishing cat at Jeypore. It was in the act of taking a four-month-old baby. The baby survived, but the cat was killed. Sterndale believed that of all the cats smaller than the leopard, the fishing cat was the only one capable of such a feat. Its strength and fortitude was amply displayed by a specimen in a zoo that broke through a partition between two cages and killed its neighbour – a full-grown leopard.

That, however, is not the end of the cat story for it may come as a big surprise to discover that domestic tabbies have featured in man-eating stories. In March 1995, for instance, the remains of a 69-year-old man was found in his house in Leiden, the Netherlands. Very little of his body was left. His 15 pet cats had eaten him!

# CHAPTER 7

---

# THE THREE BEARS

A walk in the woods, sleeping rough, or just living in the great outdoors brings people in close contact with nature, but just occasionally an adventure turns sour. A dangerous situation might have been brought on by carelessness or bad judgement, and wild animals have no respect for our mistakes and may take advantage of the opportunities presented to them. One predator that has learned to do so is the bear.

Bears generally are not man-eaters. Brown bears (*Ursus arctos*) and American black bears (*Ursus americanus*) consume little meat and very rarely go out of their way to attack and eat people. Indeed, bears tend to avoid people, only attacking when they are threatened, defending their offspring or just plain tetchy. The exception is the polar bear (*Ursus maritimus*), which is always on the lookout for a meal. Nevertheless, on rare occasions brown or grizzly bears (*Ursus arctos horribilis*) have been known to deliberately target people as prey. After all they take elk and moose calves; why not humans?

The reality is that, despite the view advanced by the popular press and motion picture industry that killer grizzlies roam the wilderness areas of North America, there are actually very few incidents in the records. Between 1900 and

1979, for example, there were only 126 recorded cases of people being injured or killed by grizzly bears in the national parks of the USA. The statistics were compiled by bear researcher Stephen Herrero, Professor of Environmental Science and Biology at the University of Calgary, Alberta, Canada. He believes that incidents across the whole of North America (including attacks outside national parks for which records are not kept) might be no more than double that figure. He also noticed that most attacks occur in the months of July, August and September, even though bears are active over a greater part of the year. Would it be naive to speculate that it is the activity of people and not bears that influences the frequency of bear attacks? More hikers, campers and picnickers are out and about during the summer, and more hunters are active in the autumn. Their increasing presence in wilderness areas would bring bears and people more closely together, and when people behave irresponsibly – poorly maintained campsites, uncleared and unsecured food and rubbish, hand-feeding and getting too close to take photographs – bears can be hazardous to human life. They might appear to be large and cuddly versions of our beloved 'teddies', but they are, in reality, big and powerful creatures.

The largest bears living today are the giant grizzly bears of Kodiak Island off the Alaskan coast and the powerful polar bears of the Arctic. Kodiak bears grow up to 9ft (2.8 m) long and weigh up to 977lb (443kg), while polar bears have a maximum length of 9.8ft (3 m) and weigh as much as 1,431lb (650kg). Eurasian brown bears, with a maximum weight of 606lb (275kg), are generally smaller than grizzlies, their size increasing geographically from west to east. The largest is the Kamchatka brown bear of eastern Russia. A full-size adult grizzly weighs as much as an adult bison, and even the smallest living bear, the sun bear at 4.6ft (1.4m) long and weighing 143lb (65kg), is still bigger than a wolf. In between these extreme sizes are the other bears, including the American and Asiatic black bears, both 5.6ft (1.7m) long and weighing up to 265lb (120kg), the spectacled bear of South America, with a length of 4.8ft (2.1m) and a maximum weight of 441lb (200kg), and southern Asia's sloth bear, which is about 4.3ft (1.9m) long and weighs 254lb (115kg). All are unpredictable and potentially dangerous, although it is the brown and black bears that mainly steal the headlines.

A large bear under threat is particularly dangerous. During a surprise encounter or when a person inadvertently comes between a mother and her cubs, the bear's natural response is to attack, and when a 450lb (204kg) bear takes on a 200lb (91kg) human, the inevitable outcome is a badly damaged person. But, generally these bears are not interested in us as food. They go for us simply because, for a startled bear, attack is the most effective means of defence. If a bear is aware of our presence it will not deliberately attack. Instead, it will do all it can to avoid us, even flee. There are exceptions, and it is these bears that are potential or actual man-eaters.

One such attack took place in the early 1900s, when an old trapper was found dead and mutilated in his cabin on the Kenai Peninsula, Alaska. The man managed to scribble a few last words, writing that he'd been 'torn up by a brown bear'. The attack was savage and horrible, but had the bear seen the old man as food?

A more recent incident, in January 1970, looked very much like a bear intent on finding and catching a meal. It took place near Fort St John in the northern part of British Columbia, Canada. A local guide and member of the Doig River Indian Band was out hunting for a grizzly. This was unusual in itself, for most bears in this part of North America would have been in hibernation at this time of the year. The hunter, however, turned dramatically into the hunted. Although there was no eyewitness, Jack Mackill, the wildlife officer who investigated the case, found clues at the scene to indicate that the guide was following bear tracks in the snow. He had been walking through forest when he passed a mossy hummock. The bear, having detected the man approaching, had hid behind the hummock and, as he passed by, circled him and then charged him from behind. He was probably killed instantly. When the body was discovered, the safety catch of the guide's rifle was still on. The body itself – partly eaten by the bear – was frozen. The bear was tracked down by helicopter and shot. An autopsy revealed that it had about 3in (7.6cm) of fat on its back and rump, so it was not starving. It had broken and loose teeth, cuts and old scar tissue on the face and head, and a torn paw, but these kinds of injuries are not unusual for a large, old male bear. Could it have been out and about, when it should have been tucked up in a cosy winter den, because of its injuries? And, with an absence of its normal diet of berries, roots and other vegetation, did it attack because it saw the guide as its only chance of feeding?

As with some cases of man-eating tigers and lions, it seems that bears sometimes (although very rarely) take to man-eating when they are old or injured. Steven Herrero, writing in *Bear Attacks: Their Causes and Avoidance*, recalls just such an incident in August 1973. The chief warden of Kluane National Park was on patrol on foot when he spotted two grizzlies. One after the other, they charged up to 20 times but he was able to fend them off with a coat and hat he was carrying. The bears, according to the warden, leapt at him, actually leaving the ground completely during the attacks. One bear grabbed the coat and the warden dropped his hat, but the bears ran off. Some time later, the warden returned to the site with a colleague when the two men noticed they were being tracked by the bears. Probably guided by smell, they found the men's camp and began to trash the place, apparently looking for food. The warden fired a shot into the air and the other man threw rocks, but instead of running away as bears would normally do, they charged. Both bears were shot and an autopsy revealed one of them to be an emaciated, old female with broken bones in the face. Her long claws indicated that she had not been digging for roots. The other bear was a sub-adult, but it was the older grizzly that had pressed home the attacks, and Herrero considered whether it was her condition that had caused her to do so. Through hunger, she had been intent on feeding on people.

The most extraordinary story of an injured bear that deliberately stalked people and took to man-eating is that of 'Old Groaner'. Its story started in 1923, when a trapper disappeared on a trip up the Unuk River, near Ketchikan. Experienced trackers went looking for him but his trail was lost at Cripple Creek. Eight years later, two gold prospectors camped there and were disturbed by the clearly audible moaning of a large, and distinctly hostile, bear. They visited the spot for several years, and each time the bear haunted their camp. On three occasions it tried to ambush the prospectors. On the third attack, one of the men was swatted by the bear's gigantic 10in (25cm) paw, but with his dog distracting the bear, he was able to let off several shots at point-blank range and shoot the beast dead; and what a strange bear it was. Its hide was hairless and as tough as leather. One of its cheek bones was completely destroyed. Part of the right side of the head was missing and the healed muscle had pulled the muzzle to one side so that its fang-like canine teeth were exposed. The other teeth were broken or badly decayed and the hinge to the right jaw was broken. It is thought that down the years the injured bear had tried to attack and feed on several people, including the old trapper in 1923, but had received several gun-shot wounds for its trouble. Five bullets were lodged above the right eye and under the jaw. Old Groaner groaned for good reason – it must have been in considerable pain and was greatly disadvantaged as a natural predator. Out of necessity, it must have fed on fish or meat infrequently and took to stalking and killing anything or anybody that was relatively slow and clumsy.

The stalking of people in this way is not the usual way grizzlies behave, but there have been reports from time to time of them doing so. On 9 September 1976, a 25-year-old man from Illinois was backpacking and camping in the Glacier Bay National Monument in Alaska. He had left a tour ship at Wolf Point on the edge of Glacier Bay and hiked to White Thunder Ridge where he made camp. Four days later, he failed to arrive at his pick-up point back at the coast and so a search party set out to find him. They found nothing.

Three days later, two hiking parties – one with four people from Seattle and another with just two people – were in the same area, camped out on the lakeside. At midday on 16 September, the four were cooking soup and hot chocolate when they noticed a bear approaching along the lake shore. They did what is recommended. They banged pots and pans and made a lot of noise but the bear just kept on coming. They gathered up their food and made a hasty retreat to some rocky cliffs nearby. From their safe vantage point, they were able to watch the bear tear their camp apart. Then it started sniffing the ground and began to follow the group's scent trail. It was heading straight for them.

They decided that caution was the better part of valour and started out around the lake, trying to shake off the bear. It kept on following, until they had completed the circuit and were back at their campsite. The bear was right behind them. They made for a hill and the bear came to within 20ft (6m) of them. They stopped and the bear was just 10ft (3m) away. Then it walked up and down, staring at them intently. They threw rocks and it retreated slightly but continued to watch. Slowly they backed away and met the other hikers at the bottom of the ridge. The bear, meanwhile, was sniffing around the place on the hill where the group of four had stopped. Both parties headed for the tour boat and reported this series of strange and disturbing events.

A park ranger was helicoptered into the area. First, he spotted a camp that had clearly been trashed by a bear. It was the camp set up by the young man who had disappeared. Not long afterwards, two people out searching for the man were guided to the camp where they were confronted by a bear. They threw rocks and it made off. They left quickly. The next day, the park ranger returned with two state troopers. They examined the campsite and found that the bear had torn up just about everything. About 150 yards (150m)) from the camp, they then found the remains of the man's body – a bare skeleton, stripped of all its flesh and body parts. Just one hand remained uneaten and the feet were still encased in boots. They also found a camera, and later developed the film. Two frames showed a young female grizzly, clearly under-nourished. The reason for this fatal bear attack can only be speculation, though hunger is the most likely cause. The young female was the only bear known to be in the area. Other bears shun White Thunder Ridge because it is poor bear country with very little to eat there.

Fatal attacks – during which bears have fed on their victims – are also not unknown in good bear country. The attacks occur mainly at camp and picnic sites, particularly places where people have not been careful to store food or garbage sensibly. The bears associate the sites with easy food and could well mistake a slumbering person for a juicy steak. The problem can be best illustrated by a series of events that took place at a national park in North America.

Between 1967 and 1980, there was a spate of campsite attacks in Glacier National Park, Montana, when six people were killed and partly eaten by bears. The attacks came as a shock to the local authorities for there had been very few fatal attacks recorded previously. The new outbreak coincided with an increase in the number of visitors and therefore garbage, and more people were entering the park on foot.

The first two attacks came on the same night, terrifying incidents that inspired Jack Olsen's *Night of the Grizzlies*. On 13 August 1967, a young man and women had hiked to Granite Park Chalet, a back-country lodge some miles from the nearest road, and put their sleeping bags down in the campground nearby. They slept in the open, with no tent. They stashed uneaten sandwiches under a log and had candy bars and chewing gum in a rucksack next to the sleeping bags. Just after midnight, a bear entered the site. The two people were woken, and pretended to be dead. But the next thing they knew is that they were cuffed by the bear and sent flying from the sleeping bags. The man was face down, and the bear approached. It began gnawing on his right shoulder. Despite the pain, the man remained still, his eyes firmly closed. It then tried the girl, returning a few minutes later to chew on the man's left arm and the tops of his legs close to the buttocks. He still did not move. Then, guests at the lodge heard the screams. The bear had returned to the girl and was dragging her away. She kept screaming, until her voice faded as she was hauled gradually further away, and she could be heard no more. She was found at daylight. She was barely alive, and tragically died later in the day. The man was very badly injured but survived.

A build-up of garbage at the chalet at that time was thought to be one reason for bears to have abandoned their normal feeding behaviour. Traditionally, they had centred on the surrounding countryside during late summer and early autumn for its abundance of huckleberries and glacier lily corms. The garbage had provided the bears with their own 'fast food take-away'. More importantly, the proximity to people eroded the bears' fear of humans. This was further enhanced by the hand-feeding of bears from the balcony of the chalet. Steven Herrero investigated the incident and concluded that 'garbage feeding and habituation' were the contributing factors which brought about these attacks.

The Glacier Park Chalet incident was not the only horror to take place that night. At Trout Lake, the fine weather and good fishing had brought in many visitors. A party of five, who worked for the same company as the two attacked at the chalet, hiked the 4 or 5 miles (6–8km) from the nearest road to the lake. They caught and cooked fish and so the smell permeated the air – an enticing odour for a hungry grizzly.

In fact, a rampaging grizzly had robbed some fishermen of their catch earlier in the summer and it had 15 or more confrontations with people at Kelly's Camp, a private piece of land within the park. It had lost its fear of people, and took to eating garbage and any other human food it could get its paws on. The day before the attack a father and son had been chased up a tree when walking a ridge near the lake. The five backpackers were unaware of the troublesome animal, and while they cooked a meal of freshly caught fish and hot dogs, it was seen approaching the campsite. The campers ran for the beach and built a fire, hoping that it would keep the bear at bay. The bear meanwhile ransacked the camp, ate the dinner and tore apart a rucksack before leaving. Afraid that it would return in the night, the campers moved to the beach and arranged their sleeping bags around the fire. They took some biscuits and hid them under the log.

At about 2 o'clock in the morning the bear reappeared. It sauntered through the first camp and then took a look at the second, stealing the biscuits under the log. In fact, it returned several times during the night, but at about 4.30 it started to sniff at the campers. It bit one young man's sleeping bag and ripped the back of his sweatshirt before he jumped up and ran for a tree. The bear was startled momentarily, and so the others made good their escape and all headed for the trees . . .

all, that is, except one. One of the girls remained in her sleeping bag. The bear had bitten on the bag and had the zipper in its mouth, so she was unable to get out. Eyewitnesses said they heard her shout that the bear had ripped her arm off, and then she yelled 'Oh God, I'm dead'. That was the last they heard. The bear dragged her body up the hill and away from the campsite. The rest of the group were terrified and would not come down from the trees until dawn. They went to the nearest ranger station and reported the attack. A day later an old female grizzly was shot and human hair was found in its stomach.

Analysing the information available, Herrero again points the finger at garbage eating, the frequency of contact with people, and the consequential lack of fear of humans. This bear, like the one at the chalet, had been seen often at garbage dumps. Today, the situation in the parks has changed. Aware of the consequences of poor rubbish disposal, the authorities and concessions have cleaned up their act, but not before there were more fatal attacks.

In September 1976, a group of female hikers were camped in two-person tents at the Many Glacier Campground. They had taken all precautions – no food in the campsite, no deodorants, and an unlocked car just 10ft (3m) away as a refuge – but during the night a bear killed one of the girls and dragged her from the tent. She was partly eaten. Some time later two young male bears were shot. They were siblings, and one had human blood on its front claws. Both had started to feed on garbage and they had learned very quickly to associate people with easy food. They had stolen food from an illegal campsite ten days earlier, their first known raid, and then ambushed hikers on the trail, breaking into their backpacks and stealing the food inside. A couple of days later they raided a garbage can at the Many Glacier camping ground, and then approached two fisherman and somebody sunbathing. The fishermen ran – one heading for a tree and the other into the water. One bear went into the water, managed to grab the fisherman by the toe, but let go when he yelled loudly. It then went for the man in the tree, and began to climb. The fisherman left the tree and joined his companion in the water. They both swam to the middle of the lake, and the bears lost interest. Three days after that, the animals raided another illegal campsite near Redrock Falls. In the course of just ten days, these two bears had changed their diet from normal bear fare to scavenging human food and rubbish. They had also overcome their fear of people and quickly learned that being aggressive towards people would gain them food. It was not long before people themselves became the food, and the bears became man-eaters.

Glacier Park was in the news again in late July 1980, when two young people were killed by a grizzly near St Mary. They had been sleeping outside their tent because the weather had been hot and humid. It was an illegal site, separate from the official campground. Their bodies were found the following morning some distance from their tent: both were partly eaten. There were no eyewitnesses, but investigators found a garbage dump, containing a putrefying horse carcass and other choice morsels, on land over which the park authorities had no jurisdiction. It is thought the bear had been heading towards the dump, when it chanced upon the two campers. It was caught and shot the next day. In its stomach were the remains of the two campers. Yet again, the conclusion must be that the presence of garbage and frequent human interaction resulted in a human tragedy.

It is Herrero's belief that habituation to humans alone – that is, without feeding on human foods – might also lead to fatal feeding attacks on people; and the sixth fatality at Glacier Park, seemed to suggest that he is right. Bear researcher Katharine McArthur Jope, from Katmai National Park, Alaska, studied the behaviour of habituated bears in national parks and concluded that as long as people remain on regular trails and travel during the day, the bears ignore them or at least are not startled when bears and people encounter each other. The same, however, may not be true at campsites.

The sixth bear victim was attacked at Elizabeth Lake in late September. The bear responsible was familiar with hikers and was not known to feed on people's picnics or from garbage dumps, yet one night it attacked somebody in their camp. The 33-year-old man was hiking and camping alone, against the advice of the park authorities. His camp had been trashed by a bear, and what was left of his body was found nearby. A couple of days later, a male grizzly was shot. The pattern of its teeth matched bite marks in a book at the camp. It was a known bear, labelled 201, which had a record of aggressive behaviour towards people around the Many Glacier area, but which had tolerated the presence of many hikers in the park. In fact, Jope had watched the bear and its behaviour towards walkers. It had learned, for instance, a kind of trail etiquette, slowing to let people and horses pass, and waiting for people to leave the trail and allow it to pass itself. In the campsite situation, however, its tolerance of people enabled it to wander freely there without fear. All it needed then was to make the association between a defenceless, sleeping human and food, and a man-eater was born.

Herrero reckons that about two-thirds of attacks on people in national parks are the result of bears either getting used to people or associating people with food, such as at garbage dumps and campsites, or both.

While the experiences at Glacier Park illustrate the problems of food and habituation, it is, of course, not the only national park in North America that has had its bear problems. On 21 July 1982, for example, a three-year-old boy was seized by a young male grizzly in a picnic area of the Banff National Park. The boy's mother and the rest of the family came to the rescue and chased the bear off before it could do any serious damage, and the child survived.

At 11 o'clock in the evening of 14 September 1998, a grizzly bit a 20-year-old woman who was working in Yosemite National Park. She was close to bear-proof food lockers at the time of the attack. Her wounds were not serious.

In May 1999, a walker out on a day's hike in Kenai National Wildlife Refuge, near his hometown of Soldotna, was killed when a bear attacked him and bit into his head. A similar fate befell a member of a seismic crew working in the Swanson River oilfield, near Kenai in February 1998.

The problem is not new. Yellowstone – the first national park in the world – and surrounding areas has had mixed success in dealing with bears. The problem started early in the park's history.

Until the practice was stopped in 1941, park authorities made the mistake of providing Yellowstone's bears with special feeding platforms where people could gawk at a variation on the chimpanzees' tea-party, only the guests were grizzly bears – wild ones. There were also open-pit garbage dumps, and open garbage cans and baskets throughout the park. Without doubt, the bears associated people with food.

Hererro has calculated that about 90 per cent of all habituated-and-garbage-related attacks during the last 100 years have been in Yellowstone's campgrounds. In the 1950s and 1960s, bears ambled without fear into camping areas at night. People were inevitably attacked. In 1967, Yellowstone closed its garbage dumps. 'Nuisance' bears, which continued to seek out human foods, were relocated or shot. Between 1968 and 1973, 200 offenders lost their lives.

Yellowstone officials are today also trying experiments with Kerilian bear dogs. They are bred to attack and drive away bears and are being used to keep bears away from certain areas, such as picnic and campsites.

Even in recent years, however, despite the measures taken to prevent people and bears coming into close and potentially dangerous contact, Yellowstone has still seen the dangers of habituation. On 24 June 1983, two men were camping in tents (although the authorities advised against it) at Rainbow Point Campground in Galatin National Forest, next to Yellowstone. They cooked a meal of steak and yams, went to a town bar for drinks and then returned to camp and went to bed. At about 2.30 a.m., they woke up with a start. Something was shaking the tent. The tent collapsed, one man screamed and was dragged out through a rip in the fabric. The other man scrambled out through the same hole. In the moonlight he could see a large bear standing over his companion. As he stood up, the bear grabbed its victim's ankle and dragged him, still screaming, about 10 yards (10m) or more. The second man grabbed a tent pole and charged bravely at the bear, yelling at the top of his voice. The bear failed to back off and he threw the tent pole at him. Being shortsighted, he raced to his tent to find his spectacles, together with the car keys and a flashlight. While he was fumbling about, the screams stopped.

**Grey wolf** © Tom Vezo, BBC Natural History Unit

Right: **Leopard** © Bernard Castelein, BI
Below: **Male and female lion pai**
© Anup Shah, BBC Natural History Unit

Above: **Snarling tiger** © Lynn M. Stone, BBC Natural History Unit
Below: **Jaguar** © Lynn M. Stone, BBC Natural History Unit

Top, left: **Young wild boar**
© Rico and Ruiz, BBC Natural History Unit

Below, left: **A polar bear investigates a tourist vehicle** © Doug Allan, BBC Natural History Unit

Left: **Grizzly bear fishing for salmon**
© Lynn M. Stone, BBC Natural History Unit

Above: **A spotted hyaena chasing off a rival clan** © Anup Shah, BBC Natural History Unit

Below: **Upright black bear in a field of dandelions** © Tom Vezo, BBC Natural History Unit

Above: **Bald eagles**
© Thomas D. Mangelsen,
BBC Natural History Unit

Right: **Male golden eagle**
© Niall Benvie, BBC Natural History Unit

Top: **Komodo dragon on Komodo Island**
© Jürgen Freund, BBC Natural History Unit

Above: **Male anaconda emerging from breeding ball**
© François Savigny, BBC Natural History Unit

**Rock python constricting a gazelle** © Peter Blackwell, BBC Natural History Unit

Other people came running. The bear was still guarding its prey about 60 yards (60m) from the campsite. Then, it made off, dragging the now limp body with it. A short while later the sheriff arrived, and a search party followed the trail of blood about 300 yards (300m) to where the body lay. About 7lb (3kg) of flesh was bitten away. Traps were set, some baited with human garbage, others with antelope. Significantly, the antelope was passed over, but the garbage attracted a large, male grizzly. It was caught and killed by injection. In its gut was human hair and flesh from the victim. The 11-year-old, 9ft (2.7m) long, delinquent bear was known to the park authorities, and was a frequent scavenger at garbage dumps. It had met with many hikers on the park's trails, but had never shown any signs of aggression. The previous month, however, it had fed on garbage and dog-food from a private residence only 200ft (61m) from the site of the attack.

Whether the bear that attacked a young boy at a Yellowstone campsite on 5 August 1984 had been exposed and became addicted to human foods we shall never know; but his parents found that all the precautions in the world are not enough to deter a determined grizzly. They had done all the right things. Their evening meal was prepared well away from the tents. Spare food was locked in the boot of the car. Pots, pans and plates were cleaned and stored away. Nevertheless, just after dark, a grizzly went on a rampage through the campsite, scattering tourists in its wake. The 12-year-old boy, who had been safely tucked into his sleeping bag by his parents, became the bear's victim. It ripped open his tent, and tried to drag him away. He struggled and the bear dropped him. The boy was unharmed.

In one attack in Yellowstone a few days previously, the actual site at which a tent was pitched for the night – not in a regular campsite, but in the back country – may have had a bearing on the tragic events that unfolded. At the end of July 1984, a Swiss woman was hiking the Astringent Creek Trail and had pitched her camp in Yellowstone. She had been warned about the dangers from bears, and in particular, about the way in which grizzlies frequent the Pelican Valley area, not far from her route. She took all precautions – placing her food away from the tent and tied high between two trees.

When park rangers found her camp, very little had been disturbed, but they noticed a tear next to the front flap of the tent. About 6ft (2m) away was her sleeping bag. A little further away, the rangers discovered pieces of her lip and scalp with tufts of hair. The woman's body, or what was left of it, for much of the soft tissue was missing, was found about 250ft (76m) away. With no eyewitnesses and no bear, despite great efforts to track the killer down, the authorities could only guess at what happened. The bear, they thought, must have been a young male, for it had climbed 12ft (3.7m) up a tree and eaten all the food suspended there. It must have been familiar with people, and would not have been apprehensive entering a camp with humans present. The key observation, however, was the realisation that the woman had unknowingly pitched her camp at a wildlife cross-roads. Not only was it close to a trail – and bears are known to use hiking trails to get about at night when walkers are not about – but it was also in an area where game passes through. It was like bedding down in the middle of the wildlife equivalent to Oxford Circus.

A similar incident occurred outside a national park in August 1974. Normally, bears outside parks have less of a chance to habituate to people – if they come close to settlements, rubbish dumps or humans, they are shot – but on this occasion a wildlife photographer was camped right on a bear trail on the bank of a salmon river. A local man who happened to be driving past noticed his tent had collapsed and reported his concern to the authorities. When Fish and Wildlife Service officers arrived they found that his ripped tent and belongings, including food, were strewn about the countryside. Nearby, they discovered the remains of his body – the head, part of the pelvis, a few ribs, shirt and pants, and a belt with a knife in a sheath.

A couple of weeks after Yellowstone's Astringent Creek Trail tragedy, another attack occurred about 20 miles (32km) away. A middle-aged couple were hiking in the park, when they were attacked and mauled by a female grizzly. Fortunately, they were able to escape with no more than cuts and puncture wounds, and managed to get back to their car.

At the time, park officials were puzzled by the attacks and those of the previous year. But, then the brain tissues of the bear that had attacked the two men at Galatin was analysed. The scientists were looking for signs of abnormalities and they found some – there were traces of the drug PCP or phencylidine, better known on the streets as 'angel dust'. In humans it can produce violent, sometimes homicidal, behaviour, but it also finds its way inadvertently into the brains of bears and other large animals. It is a component of the powerful tranquilliser used to anaesthetise bears when they are fitted with radio-collars for scientific tracking or are knocked out in order to be moved safely from one part of the park to another. Nuisance bears would have been injected with such drugs, and some of these bears – including the one that killed in the Galatin Forest – were junk-food addicts that turned to man-eating. The Galatin bear was hooked on whatever edible rubbish civilisation could provide and it was nicknamed Dumb-Dumb, because, every time it was relocated it found its way back from remote wilderness areas to the campsites and their tantalising food. It was tranquillised at least 19 times. Could there be a connection? Could residues of PCP have accumulated in its brain? The research has yet to be done.

In Europe and Asia, the grizzly's brown bear cousin has also been guilty of attacks on people. In June 1998, for example, a bear killed a 43-year-old male jogger in the forest outside the town of Ruokolahti, near the border with Russia. It was the first known fatal bear attack in Finland this century. And, exactly a year later a bear chased two fishermen on Lake Kyyjarvi, about 250 miles (402km) north of Helsinki. The bear leaped into the water and followed the fishermen's boat. A swan swooped down and began to peck the bear on the head, and the bear disappeared into the forest when friends arrived on the scene in a tractor.

In October 1998, a brown bear attacked two women picking blueberries and cranberries, near Ovgort in Siberia. One was killed, but the other survived after being taken to a hospital by neighbours.

In May 1999, a brown bear killed a man and injured two

others in Hokkaido, Japan. It was the first death of a human from a bear attack for nine years. The man was collecting plants and his family began to worry when he did not return. Then injuries were inflicted upon two women in the same forest. The same bear was thought to be responsible for all the attacks. It was tracked down and killed.

While the brown bear of Eurasia and the grizzly in North America are the biggest and most powerful bears that people may chance upon, the species more likely to be met is the North American black bear. Despite the name, black bears are not always black. There are 'cinnamon' bears living in south-west Canada, blue or glacial bears in the north-west, and pure white Kermodes bears on an island off British Columbia. Elsewhere they come in all shades of black and brown. They are smaller than grizzlies, with a body weight up to 250lb (113kg) and a shorter coat. They have one important difference – important, that is, to fleeing humans – for they can climb trees.

Despite its seemingly docile disposition, the black bear is an incredibly powerful animal. It can kill a full-grown cow with a single powerful bite to the neck and chop a branch thicker than a person's arm like snapping a twig.

There are ten times as many black bears in North America as grizzlies, which might account for statistics showing black bears making over 500 recorded attacks on people during the period 1900 to 1980. Most encounters resulted in minor injuries, but there were 35 cases with major injuries, of which 23 were fatal attacks.

Black bears are amazingly tolerant of people, no matter how moronic our behaviour. In fact, it is surprising that there are not more attacks and serious injuries considering the black bear's predilection for human foods and our inclination to play the fool. People hand-feed, pet, poke, and even shake hands with black bears, yet they are rarely provoked into attacking.

One classic story from Yellowstone recalls how a woman continually goaded a black bear into standing on its hind legs and dancing before giving it food. When the supply was finished the bear was dropping down to the ground, but accidentally caught its front claws in the woman's dress. She had minor scratches but, more significantly, the bear had exposed her bosom. Embarrassed and humiliated, she demanded the bear be killed. Fortunately, the park superintendent sided with the bear.

Although rare, many attacks in national parks come from bears looking for food. Having sussed that motor vehicles are synonymous with food – whether hidden inside or handed out from a window – black bears have been known to smash windows, rip out door and window frames, and tear away steel panels like opening a can of soup – a practice known locally as 'car clouting'. According to reports from Yosemite National Park, 1998 was a record year for such behaviour with 1,103 vehicles (six times as many as in 1993) broken into and robbed of food. Japanese compact cars, such as Hondas and Toyotas seem to be favourite targets, and mother bears appear to be teaching their youngsters how to clout. One report tells of a California resident who discovered a mother bear and cub sitting eating from a backpack he had left in the car – a Toyota Tercel. They had peeled down the right rear door. On driving the wrecked vehicle to a nearby campsite, he came across another mother bear and cub doing exactly the same thing to a parked car – again a Toyota Tercel. Another bear took a fancy to Volkswagen Beetles. Volkswagens are relatively airtight when the doors are locked and the windows wound up tight, and this enterprising black bear found a unique way to get at the inevitable cache of food inside. All it had to do was jump up and down on the car's roof until the air pressure inside burst open the doors. The bear was so adept at springing open VWs that it began to bounce on every one it could find – until it was shot.

In the wilds, the bears sometimes resort to rob-and-run tactics to ambush hikers and steal their food. Nevertheless, the number of serious or fatal injuries from black bear attacks is mercifully small. Steven Herrero records 23 people killed between 1900–80, and believes that actual predation, that is, the deliberate targeting of people as food, was the motivation for about 90 per cent of these attacks.

The black bear also differs from its cousin the grizzly in that attacks on people tend to be outside national parks. In fact, Herrero only cites one case inside a park boundary. Also, most attacks are during the day, and over half are on children. Such was the case, recorded by S.C. Whitlock, of a three-year-old girl at Mission Hill, near Sault Sainte Marie, Michigan, on 7 July 1948. She was playing outside and trying to get in through the screen door on the back porch when a bear grabbed her by the arm and dragged her to the forest. A search party with a tracker dog first found blood spots and then a shoe. The dog caught the scent and not more than 150ft (46m) away, the men found the girl's body. It lay face up at the base of an oak tree. There were deep tooth marks in the head and neck, one of which appeared deep enough to have entered the brain. The abdomen, stomach and intestines were gone, as was the flesh on the top of each leg and on one calf. On a bush nearby, a length of intestine was found. The bear was tracked down and shot. It was a healthy and apparently physically normal black bear, although it was thin and probably hungry. It was still a few weeks to the blueberry season when bears fatten up on the fruit. In its stomach were fragments of human flesh and bones.

In 1963, a failure of the blueberry crop in Alaska is thought to have been the trigger for a spate of attacks that left four people injured and one dead. Hungry black bears took to raiding campsites instead.

In the main, black bears are not responsible for nocturnal attacks at campsites, but an incursion at a site in the Rocky Mountain National Park in 1971 certainly was. In an attack reminiscent of grizzly behaviour, a black bear entered a campsite near the Holzworth Ranch at about 2 o'clock in the morning. A young couple were asleep in their mountain tent when the bear burst in. It bit the woman on the buttock and grabbed the man, pulling him about 150ft (46m) from the tent. A man in a nearby motor-home came running with a frying pan and hit the bear hard round the head. Its victim was already dead. The bear was caught and shot some time later. It was an old animal with worn teeth, and its stomach contained nothing but refuse. Just like the killer grizzlies, this black bear had been feeding at garbage sites, had become habituated to humans, and had soon turned to man-eating. On this occasion, the counter-attack by a man with a frying pan, illustrated how this species of bear, unlike the even more powerful grizzly, can be driven off, and an incident some years later shows how even a relatively small and weak person can survive by being aggressive.

On 1 September 1976, about 40 miles (64km) north of Williams Lake in British Columbia, Canada, a 10-year-old girl carrying water from a creek to her home was surprised by a black bear. She dropped the bucket, grabbed an axe lying nearby and whacked the bear twice. She ran for the front door, but the bear swiped at her, breaking her ribs, and she fell over. Nevertheless, she scrambled up and made it to the cabin, but the bear tried to get inside. With extraordinary presence of mind, the girl took a pot of boiling water from the stove and threw it in the bear's face. The bear ran away, which was just as well for clearly it was intent on feeding.

The black bear that attacked a young geologist in the Yokon-Tanana Upland area of Alaska was not so easily deterred. The story was told by the bear's victim to schoolteacher Larry Kanuit and included in his somewhat harrowing book *Alaska Bear Tales*. It was 13 August 1977, and she had been helicoptered into her mapping area. Working alone, she noticed a black bear in the vicinity and tried to frighten it away. The bear was not to be put off, however. It managed to creep up behind the young woman, swiped at her, and knocked her to the ground. It started to eat her alive, tearing the flesh from her arms, shoulder and side, then dragged her down the hill and into a ravine. Still conscious, she managed to pull out her radio and signal for help. All the time, the bear continued to feed, biting her head and scalp, until a helicopter hovered overhead and it was frightened away. After several months in hospital – undergoing surgery, skin-grafts and fitted with artificial arms – the woman miraculously survived the ordeal. A female bear was shot in the area, but it is not clear whether it was the killer. Its stomach was filled with blueberries, but also contained semi-digested substances that could have been human flesh or clothing.

This attack was a rare example of a black bear seeing people as prey, and a similar incident occurred a couple of years later in Mt Robson Provincial Park, British Columbia. On 4 July 1979, a woman was hiking alone when she spotted a bear. It did not attack immediately, but when it did the woman played dead. Offering the bear an arm to chew on, in order to protect her more vital organs, she waited for help to turn up on the trail. When a party of people arrived, the bear took off.

In the summer of 1980, another confrontation ended tragically not for one, but for two victims of a bear attack. The location was an oil installation near Zama, northern Alberta. The geologist for the camp went rock-hunting along a stream bed, but failed to return. Two of his colleagues – a man and woman – went to look for him, but found a black bear instead. The pair ran for the nearest tree, but the woman tripped and fell. She then panicked and could not climb. The man reached down and tried to help her, but she slipped from his grasp as the bear took a lunge at her. When she landed on the ground, the bear seized her by the neck and shook her hard. Her neck broke instantly, and it dragged her lifeless body into the bushes. The man remained in the tree, but the bear came back for him, climbing about 12ft (3.7m) up the trunk. The man kicked it in the head and pushed it with branches, and it clambered back down, only to reappear and climb back up again and again. Eventually, another workmate came by and ran to raise the alarm. Returning with a gun, he was spotted by the bear and it charged. He took one shot, the bear screamed and ran away. Sometime later, the bodies of the girl and the geologist were discovered. Both had been partly eaten. A male black bear was killed in the area the following day.

Interpreting the events that led to the deaths of the two oil workers, Stephen Herrero considers that the bear undoubtedly looked at people as food, and that it would have killed anyone in sight in a kind of 'feeding frenzy' known as surplus killing. The Alberta affair, however, was nothing compared to what was to come.

On 13 May 1978, four teenage boys were on a fishing trip in Algonquin Park, Ontario. They arrived from the Canadian Airforce Base at Petawawa in the morning and tried several creeks for speckled trout. One of them caught four. With his luck that day, he decided to fish a little more, in the late afternoon. Two of the others went to look for him, and the fourth boy, an 18-year-old and the oldest of the group, remained behind in the car where he caught up on some sleep. An hour later the eldest boy woke, but the others were nowhere to be seen. He shouted, honked the horn and finally drove around looking for them but they had simply disappeared. He drove home and raised the alarm. Soon a search party was assembled, and the bodies of the three boys were found. A black bear was standing guard over them. The bear was shot.

With no eyewitnesses, the actual sequence of events is not clear, but investigators at the scene of the killings put together a likely scenario. The boy who had been fishing was attacked from behind. Broken branches indicated that he probably put up a fight, but the bear killed him and dragged his body upstream. When the other boys arrived the bear ambushed them, and broke their necks. All three bodies were dragged into what amounted to a caching site. It was probably another case of surplus killing.

Other incidents have a similar feel to them. In May 1983, a 55-year-old fisherman was killed and partly eaten by a black bear in central Saskatchewan. Two fisherman in the area were also attacked but one managed to stab and kill their attacker with a filleting knife. In the same month a trapper's body was found. He had head and neck injuries matching those seen on other black bear kills.

A third fatality in 1983 implied a new black bear attack tactic. A 12-year-old boy was with a group of people camped beside Lake Canimina in La Verendrye Wildlife Reserve, Quebec. He was grabbed by a bear and dragged about 100ft (31m) from his tent. The bear had visited the camp three times during the night, on one occasion breaking into the camp larder suspended from trees about 165ft (50m) from the campsite. On its final visit it killed the boy. According to Herrero, this was a rare event.

Similarly, a 14-year-old boy was sleeping at a Boy Scout camp at Tomahawk Bay, Wisconsin, when he was attacked and seriously injured by a black bear. The incident occurred in August 1999.

A mother from Texas and a man who tried to rescue her were killed by a black bear at the Liard River Hot Springs Provincial Park in British Columbia. The woman's 13-year-old son tried to help, as did a 20-year-old student from Calgary who was conducting research in the park, and both were mauled but survived. Hikers who chanced on the mayhem threw rocks and sticks in attempts to distract the bear and stop the attack but they were to no avail. Eventually, a tourist with a gun arrived and shot the bear dead, but it was too late for the two victims.

In June 1998, a woman walking her dogs was attacked and seriously injured by a black bear in woods near Hampton, New Brunswick. She was badly slashed on the face, arms and torso but was able to get away from the bear and run a half-mile or more to a phone where she called for help.

One of the most bizarre black bear attacks, however, must have been the one that took place in Alaska. A 55-year-old trapper, who lived in a cabin about 100 miles (161km) north-west of Anchorage, Alaska, was on the receiving end of a predatory black bear. The bear broke into the house, but the trapper grabbed a rifle and made a quick exit through a window. Circling around to the front door, he shot the bear trying to escape but it mauled him. The bear died just outside the door, and the trapper made it to his bed but collapsed and died too.

The American black bear's equivalent in the Old World is the Asiatic black bear (*Selenarctos thibetanus*). It is mainly a herbivore but will take to eating carrion. Under certain circumstances it comes into conflict with people. In October 1986, for example, black bears left the snow-bound forests in the Kashmir Valley and raided small villages. Scarcity of food was blamed for attacks on people and livestock. Three people were killed and 29 others badly mauled during an eight-day spree.

The most awesome of the bears, though, must be the magnificent ice bear. The polar bear (*Ursus maritimus*) vies with the Kodiak bear for the title of world's largest bear. Being a solitary migrant, it is always on the lookout for food, and humans are as fair game as its more usual prey, the ringed seal.

Until mineral exploration and exploitation was widespread in the Arctic, encounters between people and polar bears were relatively rare. The local Arctic people knew about its ways and respected it, and very few were hurt. Then came oil, and the fatalities due to polar bear attacks increased dramatically. On an artificial island in the Beaufort Sea, for example, a construction worker who was working alone on the deck of a barge was surprised by a polar bear and killed. The bear dragged his body on to the sea ice and started to eat him. A search party discovered the bear and the partially consumed body, and the bear was shot. It was found to be a young, emaciated bear which had probably taken advantage of an easy meal.

In August 1975, two workers at a geological survey camp on Somerset Island were attacked and mauled by a polar bear. One was dragged from his tent, but a geology student was quick enough to shoot the bear dead before the bear made its kill. The two survived. Four days later, Canadian geologists working in the High Arctic heard a snuffling noise outside their tent. The men listened, and something pushed against the canvas. It was a polar bear. Adopting the method used to break into ringed seal dens to get at the occupants, the bear stood on its hind legs and lunged forward with its huge forepaws. The tent held, so it tried ripping the canvas flaps and biting the steel supports. The men's rifle was outside. They cut a hole in the side of the tent in order to fetch the gun, but the polar bear poked its head in. One man grabbed a piece of meat and threw it as far as he could, and the bear left them to search for it. The gun was unloaded but three bullets were rammed home hastily and the bear was shot. The two men trekked 4 miles (6.4km) to the next camp. They survived, but an oil worker attacked by a polar bear earlier the same year was not so fortunate. He was seized in the darkness of the early morning and his partly eaten body was not discovered until later in the day after fellow workers had followed a trail of blood across the ice.

Polar bears, though, would not be too excited about people as food. We do not have enough fat on us. The bears normally eat the blubber from seals – their natural prey, leaving much of the meat to the Arctic foxes that often accompany them at kills. So, a polar bear that sees humans as serious food is likely to be a very hungry one. In the vast Arctic wilderness, food is hard to come by and so a polar bear will take a closer look at anything animate. Unlike its two cousins, the grizzly and black bear, it can usually be persuaded to push off by shooting into the air or letting off thunderflashes. But at certain times of the year it is driven by hunger to be more persistent.

At Churchill, Manitoba – the undisputed polar bear capital of the world – bears arrive each autumn. The town is on a polar bear migration route. The bears are on their way from spending the summer in the forests to Hudson Bay where the seals are waiting for them. If the formation of ice is late and the bears are prevented from hunting, Churchill has a problem, for the hungry bears roam the streets in search of food.

Writer Ed Struzik recalls a night when all hell broke loose in Churchill. He was visiting one of the biologists stationed there when a man in the street started screaming. Drawing back the curtains, Struzik was horrified to see a man with his head firmly caught in a polar bear's mouth. He was being shaken around like a rag doll. People began to emerge from their homes and, dressed only in pyjamas and night-shirts and armed only with broomsticks and insults, they tried to chase it away. It charged several of them, and was only brought to a standstill by a shot from a rifle. A police investigation revealed that the man had been drunk and was looting amongst the ruins of a burned-out hotel. The polar bear had had the same idea.

The bears' favourite site, however, is inevitably the rubbish dump and nowadays, wildlife officers are lying in wait ready to capture and relocate the bears to a safe distance from the town. Churchill has learned to live with its seasonal invasion. It is also learning how to deal with the seasonal influx of tourists who swamp the town each year. Despite this, confrontation between bears and people is minimal.

The same problem exists elsewhere, however, without a full appreciation of why the bears are there at all. The warnings of wildlife officers have often been ignored, and mining camps, such as the Polaris lead-zinc mine on Little Cornwallis Island, have been set up on traditional polar bear migration routes. Bears have an extraordinary sense of smell, and so the plume of breakfast smells or the aroma of rotting garbage downwind of a camp is like a magnet to any passing bear. Faced with the clatter of machinery and other loud noises, a bear is probably reluctant to enter a camp during the day, but come the quiet and darkness of early morning, it will slip in silently and unnoticed. And woe betide anybody who is out and about or asleep in a relatively flimsy tent.

Younger bears can be a problem. They are abandoned by their mothers and are not fearful of people. They are not so adept at hunting and are often hungry, so they are not easily put off. In August 1998, a 17-strong party of students from Durham University discovered this for themselves. They were exploring the Svalbard Islands when a polar bear came close to their camp in the Hornsund National Park. They were able to drive it away, but it came back and broke into the mess tent. The people banged pots and pans, blew whistles, let off thunderflashes and flares, and fired shots into the air, but when the bear cornered two men they were forced to shoot it. The day after, another bear turned up and behaved in the same way. It attacked one of the expedition members and it too was shot. 'If I had missed,' said the expedition leader, 'the person the bear was charging would have died, along with several others in the party.'

A few days previously another team in the same area was attacked by a polar bear, and again it could only be stopped when it was shot. And, on the sea nearby, an unarmed team of Irish explorers were on board a small boat trapped in the ice and were surrounded by several bears. They only had flares to scare the bears away.

In July 1999, one person was killed and two other were badly injured when a large polar bear attacked their remote camp at Corbett Inlet on the western shore of Hudson Bay in the Canadian Arctic. And in August 1999, two Norwegian adventurers were stranded on an Arctic island for 12 days, after a 700 mile (917km) journey around the Svalbard archipelago when a polar bear attacked them and ripped up their kayaks.

Despite our ignorance, however, fatalities due to polar bear attacks are relatively few. The increased demand for more adventurous holidays as gradually more people discover the thrill of Nordic pursuits, such as cross-country skiing and white-water canoeing, may just see an increase in attacks.

# CHAPTER 8

---

# HYENAS, PIGS AND EAGLES

While there are bears in Europe, South America and Asia, there are no bears living today on the African continent. There is, however, a mysterious creature for which there have been sightings which strongly resembles a bear – it is the so-called Nandi bear.

The Nandi bear is a crypto-beast reported to kill and eat people throughout East Africa. Also known as the *chemosit* (meaning 'the devil') and *geteit* (meaning 'brain-eater'), depending on the district in which it was seen, mere mention of the Nandi bear is greeted with fear by local people. It is supposed to be a nocturnal, bear-like animal that stands about 5ft (1.5m) tall, and is very dangerous. Geoffrey Williams mentions it in an edition of the *Journal of the East Africa and Uganda Natural History Society* published in 1912. He was travelling on the Uasingishu. He had reached the Mataye and was walking on foot towards the Sirgoit Rock when he saw a strange animal sitting on its haunches not more than 30 yards (30m) away. Its forequarters and legs were thickly furred but the hindquarters were bare. It resembled a giant cross between a bear, baboon and hyena.

In fact, a giant baboon – of which an extinct species once vied with early man for dominance of African plains – is a possible candidate for the identity of the Nandi bear. After all, baboons, drills and mandrills have acquired a reputation for killing children. Colonel C. R. S. Pitman, writing in his *A Game Warden among his Charges*, published in 1931, tells of five local children being attacked and two killed by baboons. Giant baboon or not, local people warned Williams of the beast. They said it was dangerous to sleep with the tent flaps open for it attacks in the night. Solitary walkers are most at risk, they said.

Investigating the literature, cryptozoologist Bernard Heuvelmans discovered many references to the Nandi bear and devotes an entire chapter to the phenomenon in his *On the Track of Unknown Animals*. Such was its reputation, he discovered, entire villages would up sticks and move out of the area if the Nandi bear was about. It was said to attack people on sight, rip the tops of their head off, and eat only their brain.

In 1925, one village was so fearful of the Nandi bear that the elders asked the Kenya government to help them. The last straw was when the creature forced its way through a thick thorny fence, cut its way into the wall of a mud hut, and carried off a six-year-old girl – behaviour reminiscent of the Tsavo lions.

Captain William Hitchens of the East African Intelligence and Administrative Services was sent to investigate. Along with a group of tribesmen, Hitchens beat the brush around a nearby kopje (boulder hill) and put up all manner of animals, but no Nandi bear. Then, in the night, something came and rattled his tent so much that the whole thing collapsed and his small dog, attached to one of the poles, was snatched away. What Hitchens recalls most vividly, though, was the horrific howl that cut through the night. In the morning, he and his companions followed a trail of blood spots that must have been from his dog, and came across tracks which led across the kopje and into the forest. But they never found the Nandi bear.

Since then, many reports of sightings of the beast have come from travellers, settlers and government officials but there is still no specimen, photograph or film. For the moment, the Nandi bear, with its brain-eating habits and blood-curdling scream, will have to remain in the unsolved files of cryptozoology.

A possible candidate for the identity of the Nandi bear is a giant hyena. In the 1920s, unusually large specimens of spotted hyenas were thought to be responsible for human deaths in the district of Karamoja, where an old woman was dragged from a hut and eaten and a seven-year-old boy asleep in thorn scrub was taken and devoured.

The hyena is a wily predator more than capable of attacking a person who is very young, injured or alone in the bush. It has occasionally attacked and killed people when they are sleeping, biting the face with bone-crushing teeth and strong jaws. In fact, it has a set of the most powerful jaws of all the carnivores, and is quite capable of cracking and splintering bone. At one time, it was considered a cowardly scavenger, but it is now known to be an active predator in its own right. Although superficially dog-like, the hyena is more closely related to the civets. It lives in clans, sometimes containing more than 50 members, and hunts co-operatively with hunting parties of 10–15 hyenas, much like wolves, wild dogs and lions. When prey is brought down, the pack is quick to tear the corpse asunder. On one occasion in East Africa, a large zebra was seen to be eaten in just 15 minutes, and few scraps were left. On another in South Africa, a 221lb (100kg) wildebeest was ripped apart and consumed in just ten minutes.

Of the four principal species of hyenas, and their relative the aardwolf (*Proteles cristatus*), the spotted (*Crocuta crocuta*) and striped (*Hyaena hyaena*) hyenas seem to be potentially dangerous to people. Indeed, none other than Teddy Roosevelt wrote about the way hyenas 'enter native huts and carry away children or even sleeping adults'.

They became front-page news in 1908 and again in 1909, when they took advantage of the bonanza of human corpses during epidemics of sleeping sickness in Uganda. Not content with the dead, they became much bolder and also dragged the sick and helpless from their huts. Armed guards were posted at the sleeping sickness camps. Eventually, they attacked healthy people too. Sir Alfred Pease tells how a hyena grabbed a local hunter and tore away part of his face. Not far away, in what was once Rhodesia, the administrator of the north-west part of the country, Major R. T. Coryndon, was attacked in his bed.

In 1910, a hyena carried off a man on the outskirts of Nairobi, and some years later, the four-year-old son of a Baptist minister was dragged from his parents' tent in the Kruger National Park, South Africa. The boy's screams brought his father running to the rescue, but the hyena was 30 yards (30m) into the scrub before dropping the boy.

In 1950 and 1951, people were attacked at all times of the day in Manyoni. The hyenas tracked and chased humans just like other prey, taking it in turns to hustle women and children. The victims were 'devoured limb by limb' during the chase.

In 1967, the drought in Turkana, East Africa, forced packs of hyenas to raid villages where they pulled down both cattle and people. The local human population got their own back by eating the hyenas they had speared.

Even today, there are many stories of spotted hyenas entering tents and open doors at night, particularly if there is an enticing smell coming from inside. They will also attack people in tents without warning. Where a lion might circle a tent, sniff about out of curiosity and leave, the hyena will rush in and take a bite out of anything in its path, including any exposed part of a person. There are many cases of Africans with half their faces bitten away, the result of hyena attacks.

Hyena biologist Ronald Tilson recalls several occasions when hyena paw prints were found outside his tent. The animals had visited his study camp every third or fourth night without rousing anybody. On one night, however, Tilson woke up to find himself face to face with a large spotted hyena. 'Its breath', he recalls, 'was unimaginably foul.' Though his mouth was dry and he could not yell or scream, he managed to hiss at the intruder and it eventually backed away and left. The brief but terrifying encounter left a lasting impression on Tilson, but had it simply been curious, he wondered, or would it have attacked if he had not woken up?

In Malawi, there was a time when spotted hyenas specifically targeted humans, attacking during the day. In one part of the country a spate of raids saw 27 people killed during a six-year period. And in southern Sudan, a region plagued with civil war for many years, reports emerged that told of Dinka tribespeople being attacked by hyenas in 1989. The animals entered villages, such as Yerol which is about 600 miles (966km) south of Khartoum, and took children, the elderly and sick. Here, they would prowl the streets at night, entering huts and dragging out sleeping children and sick adults. They even leaped through the windows of the village 'hospital', biting the patients on the face. They also took bodies from the mortuary. This predilection for human flesh began three years previously when over 3,000 corpses were left unburied after a battle between government and rebel soldiers. The hyenas made light work of the bodies.

In some parts of India, where tigers, leopards, wolves and feral dogs are thought to account for many deaths, the human population must contend with hyenas too. Here, the striped hyena is a recognised man-eater. In May 1962, for example, nine children from the town of Bhagalpur in the Bihar district were thought to have been taken by hyenas in just six weeks.

Of the land-based mammals, it is the large and fearsome hunters – the hyena, big cats, and wolves – that have been implicated in episodes of man-eating, but there have been instances when animals not known traditionally as predators, including some of the smaller mammals, have had a go at us.

An unexpected man-eater is the pig. Wild boar (*Sus scrofa*) are known as wild hogs in North America, but wherever they are found they are quite capable of slicing up a human body. They probably fed alongside wolves and other scavengers on the corpses at battlefields. There are few recorded cases of wild pigs killing people, but as any hunter will point out, wild pigs leave no evidence behind: they eat everything – flesh, skin, bones, the lot.

There is one case quoted by Don Zaidle in his *American Man-Killers* of a 79-year-old man from southern Texas who was charged by a dozen wild hogs. All he had to defend himself was a wooden stick. The hogs made light work of that, splintering the wood with one bite of their tusk-filled mouths, and then turned to the man himself. One bit into his thigh while the other savaged his ankle. The man fell and the group gathered round to feed, a large boar taking a chunk out of his leg. Summoning all his strength, the old man punched out at a hog. Miraculously, it screamed and ran, the other hogs following it back into the brush. The man hobbled into his house and was just able to phone a neighbour for help when he passed out.

Brown rats (*Rattus norvegicus*) and house mice (*Mus musculus*) have acquired an evil reputation down the centuries, but it is often the fantasy more usually found in myth and legend. There is the legend of Bishop Hatto, for instance, who lived in the 10th century. He was locked up in the Mauseturm in Bingen where his body was stripped to the bone by mice. There are records of rats biting an exposed nose, finger or toe but never eating a person 'down to the bone'. In September 1983, however, workers at the highest cable-car site in Europe at Saas Fee, Switzerland, were attacked and bitten by what was claimed to be 'starving' rats.

Rats are primarily seed eaters and secondarily scavengers. They would not deliberately set out to feast on human flesh, but if it should be available in the form of a helpless baby in its cot or a corpse on a battlefield, like any opportunistic scavenger, they will gnaw away at anything they might find. This happened in October 1983, when a cancer sufferer living alone in a council house in Ilkeston, Derbyshire, died and, during the three weeks before he was found, his body was partially eaten by rats.

Roger Caras, in his book *Dangerous to Man*, cites the case of a man and his wife who were woken up in their Manhattan upper East Side apartment by the cries of their 18-month-old son. When they went to him, they discovered he had several bites on his cheek, ear and neck. Next to him on the pillow was a very large rat. The father took a broom and tried to hit the rat, which attacked and bit him deeply on the leg and then turned to attack his wife. All the family were hospitalised, and a local judge ordered exterminators into the building. They caught 43 rats on their first try.

In July 1998, a six-month-old boy was in critical condition after being bitten by a pet rat at his home in Arlington, Texas. When his mother found him, his face was covered in blood. His face, right arm and left foot were badly damaged and hospital authorities considered reconstructive surgery to rebuild the muscles of his upper left arm. The white rat had escaped from a fish tank.

Of the other small mammals, the domestic ferret (*Mustela putorius furo*) has been known to take a chunk out of an exposed body part. Like rats, it can do a baby much harm. A young lady from Cornwall, writing to the popular British periodical *Country Living*, states that she was 'nearly eaten alive by a ferret as a baby'. Fortunately she was saved from serious disfigurement, but she still had 100 stitches to her face and nearly lost her left eye. In the USA in February 1998, a pet ferret bit a five-week-old baby girl on the face and chest at least 50 times. The baby had been sleeping between her parents.

Human babies are vulnerable also to birds of prey. There have been many sensational stories, and most prove to be folk tales passed from generation to generation without any basis of fact. There are a few, however, that appear to have some semblance of truth.

Large eagles are potentially dangerous because of their ability to carry relatively heavy loads. The rule of thumb appears to be that some of the larger species of eagles, such as the golden eagle (*Aquila chrysaetos*), white-tailed sea eagle (*Haliaeetus albicilla*), Steller's sea eagle (*Haliaeetus pelagicus*) and bald eagle (*Haliaeetus leucocephalus*) can haul their own weight aloft. Indeed there is a reliable report of a bald eagle having dropped a 15lb (6.8kg) mule deer fawn after hunters yelled at it; and in *I Was a Headhunter*, Lewis Cummings reports having seen a large South American eagle, probably a harpy eagle (*Harpiu harpyga*) carrying a half-grown deer weighing an estimated 35lb (16kg). With its large wingspan and aided by an up-draught of air or gust of wind blowing in the right direction, an adult eagle might just be able to struggle into the air with a human baby or infant, and there is at least one baby-snatching incident, described in *Living Wonders* by John Michell and Robert Rickard, that appears to really have happened.

It took place at Leka, a small village to the north of Trondheim in Norway. Svanhild Hansen was four years old at the time, but quite small for her age. One day she was playing in the yard of her parent's farmhouse when a white-tailed sea eagle swooped down and grabbed her. These are big, vulture-like eagles, with a huge deep bill and a long neck. They have a wingspan up to 8ft (2.4m) across. They more usually eat fish, birds and carrion, but have been known to take sheep, reindeer and red deer calves. On this occasion human flesh was nearly on the menu.

Helped by an up-draught of wind, the eagle lifted Svanhild off the ground and, carrying her in his talons, flew the mile or so to its eyrie about 800ft (244m) up on a mountain cliff. It failed to reach the nest with its heavy load, and dropped the girl on to a ledge about 50ft (15m) below. Svanhild's parents were understandably frantic, and a search party from the village was immediately gathered together. Guided by the eagle, which was circling continuously over the high ledge, the villagers eventually reached the girl. She was not hurt except for some minor scratches and bruises and so they were able to bring her down safely. Svanhild still has the dress she wore that day and the holes made by the eagle's talons are still visible.

Similar events are said to have occurred down the centuries. In the *Description of the Western Isles of Scotland*, published in 1716, there is the testament of a young man from Skye who, as an infant, was left by his mother near his house on the north side of Loch Portrie. An eagle came and carried him away, dropping him unhurt on the south side of the loch. Some shepherds heard his cries and took him back to his mother.

Further to the north, on the Hebridean island of Uist, a couple were helping with the corn harvest while their baby girl was left at the edge of the field. The year was 1790. The parents heard a loud flapping noise and turned to see a sea eagle trying to take their baby. All the workers gave chase but the bird flew out with the baby over the sea, towards an island on which sea eagles were known to nest at that time (white-tail sea eagles became locally extinct but were re-introduced successfully to some Scottish west coast islands in recent years). Several of the men took a boat and rowed the 3 miles (4.8km) to the island. With the help of crofters living there, they were able to lower a boy down to a nest on the cliff side. In it they could see the baby being pecked at by eaglets. The boy was able to carry her safely from the nest and the two of them were hauled to the top of the cliff. When she grew up, the boy married her and their descendants live in the area to this day.

In 1868, in Tippah county, Missouri, a local teacher described how eagles had been a particular nuisance, carrying off piglets and lambs. One Thursday morning, however, one grabbed an eight-year-old boy playing near the schoolhouse and flew off with him. The other children screamed and the teacher came running but to no avail. The eagle dropped the boy and he was killed on hitting the ground.

In *Nature Parade*, published in 1947, Frank Lane tells the story of a young Irish girl who was similarly carried across the Kenmare River by a golden eagle. And in a newspaper story in London's *Evening News*, dated 4 November 1978, Lee Wilson writes of two abductions. At Ammasia, southern Anatolia in 1937, a mother was washing clothes on the river bank and an eagle carried away her baby. In February 1953, the same thing happened near Damascus. Fortunately, the Syrian baby was found alive and well in an eagle's eyrie on the edge of a cliff. For some reason, the three eaglets in the nest had not attacked it.

Circumstantial evidence seems to point to an avian abduction in the European Alps. In 1950, a mountain guide discovered a child's skeleton high in the French Alps. It was thought to be that of a four-year-old boy who had disappeared three years previously. The child is unlikely to have climbed to the crag, and so it is believed he was carried there by a large eagle. It would have not been the first time. A similar incident occurred in the French Alps in 1838. A five-year-old girl was playing with a friend on an alpine meadow when an eagle swooped down and carried her off. Some people nearby heard the scream of her friend and searched for the girl. All they found was her shoe on the edge of a ravine. Her mutilated body was not discovered until two months later.

Eagles are not the only large birds implicated in attacks on children. Vultures – both New and Old World species – are suspect too. On 25 July 1977, two large black birds with 8ft (2.4m) wingspans and white rings around the neck – a description that matches the condor – attacked a bunch of children playing in a backyard at Lawndale, Illinois. One grabbed a 10-year-old boy and carried him about 2–3ft (1m) from the ground a distance of 20ft (6.1m) before the boy hit the bird with his hand and it dropped him to the ground. The two birds flew off in the direction of Kickapoo Creek in central Illinois. Several people saw and confirmed the description of the birds, but local conservation officials pointed out that no birds capable of lifting a boy off the floor live in that part of America. Understandably, the case was controversial, but the boy's parents stuck by the story. Not long afterwards, in the same area, farmers saw a large bird grab two piglets, each weighing 20lb (9.1kg), and carry them away, and a truck driver saw a bird attempt but fail to pick up a piglet.

In the Old World, too, it seems vultures have augmented their more usual scavenging habit with a bit of baby-snatching. On 12 July 1763, a peasant couple in the Bernese Oberland left their three-year-old daughter asleep near a stream while they worked nearby. As lunchtime approached the parents returned to the spot where they left the baby girl and found she had disappeared. They searched the hillside but could not find her. In the meantime, a man was walking on the other side of the hill and heard a child crying. As he approached the source of the sound a lammergeier or bearded vulture (*Gypaetus barbatus*) flew up in alarm. On the ground was the child with a badly injured arm, but still alive. In fact, she lived to a ripe old age and was known in the town of Gewalswyl where she lived as 'Lammageier Anni'.

In certain parts of southern Asia, vultures eat people. They do not attack or kill living persons, but scavenge on dead human bodies. The Parsees, who fled Persia for India in

AD636, do not cremate or bury their dead. With a respect for the elements – fire, earth, air and water – their dead are left exposed in special roofless buildings, to which non-Parsees are not admitted. The most famous are the Towers of Silence on the Malabar Hill in Bombay, built in 1672 by Seth Modi Hirji Vachha. Here, even today, 8,000 year-old traditions are still followed. The corpse is carried by special bearers into the high-walled, open-topped buildings. It not only decays naturally in the sun and rain, but is also helped on its way by flesh-eating birds such as vultures. As the body is put on its place of rest, the birds are seen to gather around the walls, like hunch-shouldered omens of death. When the funeral party has withdrawn and the body is left alone, the birds drop in to feast on the human flesh. Meanwhile, the relatives prey and perform ceremonies in special gardens, where there are flowers, fountains and peacocks. They continue for the following three days, while the dead person's soul is awaiting judgement.

Similar rituals are performed in Buddhist Tibet. Here, the body is carried by a professional corpse-carrier to a prominent rocky point outside the village. The relatives dismember the body, breaking the bones so that the vultures and other birds can consume it entirely. The local belief is that the body has no significance without its soul, and this way of disposing of it ensures there is nothing left.

# CHAPTER 9

---

# MAN-EATING CROCODILES

While travelling down the River Nile, Sir Winston Churchill once noted, 'I avow, with what regret may be necessary, an active hatred of these brutes and a desire to kill them.' Churchill's venom on this occasion was directed at the crocodile, and the formidable Nile crocodile (*Crocodylus niloticus*) in particular. Despite its common name, it is found in many lakes and rivers throughout Africa, including the Nile itself and even in the dwindling lakes of the Sahara. It is also present on the island of Madagascar where one population spends the dry season hiding in the Ankarana caves. Although mainly found in freshwater lakes and rivers, it is also at home in freshwater marshes, mangrove swamps and estuaries. Wherever it lives, it eats just about anything it can seize, including youngsters of its own kind.

Its hunting ground is at the water's edge. Large animals are grabbed when they come to drink or are attempting to cross. Indeed, crocodiles are bright enough, despite a brain the size of a cigar, to learn how to anticipate a meal by watching the behaviour of potential prey. They recognise when animals come to the river to drink at roughly the same time each day. The attack strategy is common to most crocodile species. If the crocodile is basking on the river bank and spots potential prey, it quietly slides into the water and moves slowly unseen below the surface towards its victim.

Although a large and lumbering animal with stubby legs, the Nile crocodile has an amazing turn of speed over short distances, but stealth is its main weapon. It lies motionless in the water, and, when the prey is within range, it thrusts forwards at high speed, using its powerful tail for propulsion, and grabs the head or a limb. It drags the surprised victim under the water where it drowns. It can do this because it uses oxygen at a relatively slow rate and the haemoglobin in the crocodile's blood has a special property – bicarbonate ions that bind to haemoglobin which releases more oxygen to the tissues – that enables it to remain below for over an hour or more.

Once the prey is subdued, large animals are dismembered and disembowelled in a characteristic way. The crocodile cannot swallow a large animal whole so it grasps a piece of meat in its jaws and then rotates its entire body – the so-called death roll. This enables it to twist off the chosen morsel, which it then downs in one or two big gulps. At times, humans are consumed.

One of the legendary Nile crocodiles was 'Kwena' in the Okovango Swamp, Botswana. Kwena was responsible for many human deaths until it was killed by hunter Bobby Willmot in 1968. The croc was over 19ft (5.8m) long, and weighed an estimated 1,800lb (816kg). When its stomach was sliced open out fell two goats, half a donkey and the clothed trunk of a native woman.

Indeed, the Nile crocodile is probably the most abundant large predator that people are likely to encounter in Africa. People tend to settle alongside rivers and so the crocodile has always been an uneasy neighbour. In days gone by, when much of Africa was wilderness, the rivers and lakes must have been swarming with crocodiles and there would have been many human deaths. Estimates of 20,000 per year have been suggested in the past, and one crocodile living in the Kihange River, Central Africa, alone was credited with 400 human victims. The early explorers came back with lurid tales of people being eaten alive.

In 1862 Sir Samuel Baker took a large expedition consisting of many river craft up the Nile in order to explore southern Sudan on behalf of the Egyptian Government. It lost many people to crocodiles, and others were maimed by their attacks. One man, who was collecting water plants at the edge of the river, had his arm bitten off at the elbow. His companions were able to pull him free. Another was grabbed by the foot in shallow water while pushing a boat off a sandbank. He was also saved, but still lost his leg. A crewman dangling his feet in the water was seized by a crocodile and dragged under: a hundred or so onlookers were powerless to help. A crocodile even took possession of a dock cut in the river bank which was thought to be safe to bathe in. One evening a soldier went there, but all that was to be found on the dockside was some clothes and a red fez. Other soldiers jumped into the water to search for him. His mangled body was discovered on the bottom. Some time later, another crewman was taken: he was swimming in the narrow canal between the dock and the river.

Elsewhere in Africa the story has been the same. On New Year's Day 1896, English big-game hunter Arthur H. Neumann was bathing in the River Omo at Kere. His Swahili servant Shebane brought his chair and towels. Neumann dried and dressed himself and was just tying his shoe-laces when he heard a cry. Turning towards the river he saw Shebane held in the jaws of a large crocodile 'like a fish in a heron's beak'. The servant was pulled below and disappeared before anybody in the camp could save him.

In *Crocodiles: Their Natural History, Folklore and Conservation*, C. A. W. Guggisberg tells of another harrowing tale from a medical officer travelling on the Congo River. As his steamer slowed down to manoeuvre around a sandbank, he saw a man paddling his canoe followed closely by a crocodile. The canoeist spotted the crocodile and high-tailed it to a small island. He left his boat at the water's edge and ran up the beach. Eyewitnesses on the steamer gave an audible sigh of relief, but this quickly turned to one of alarm when they saw the crocodile leave the water and run rapidly after the boy. He reached the end of the island and, obviously terrified, had to double back. The croc swiped at him with its tail, knocking him over. It grabbed him and dragged him into the water: he was not seen again. Nobody on the steamer had a rifle, and so they could do nothing but watch in horror.

Crocodile attacks on land are less common, yet not unknown. Major A. St H. Gibbons, travelling in Barotseland, reports an account of a crocodile that entered a village hut at night and grabbed a man who had been sleeping inside. The victim was dragged to a nearby river.

The stories of attacks in water are legion. Often the crocodile does not consume its victim immediately, but hides it. At Mzima Springs, in Tsavo National Park, a young girl watched her brother being seized by a crocodile. His body was found later in the reeds. And J. Stevenson-Hamilton tells of a Bantu youth who was taken while fetching water from the Sabi River. Villagers with spears probed the river and his body was discovered below a rocky overhang in the river bank.

Although there are few official records, it has been estimated that in more recent times over 1,000 people a year succumbed to crocodile attacks. Particular reaches of some African rivers, such as parts of the Juba River and the lower parts of the murky Pongola River, once gained notoriety. So has Lake Rudolph, in Kenya. In Alistair Graham's and Peter Beard's book *Eyelids of the Morning*, they tell of an American peace corps worker who was taken in the muddy Baro River that feeds Lake Rudolph. His colleagues witnessed the attack, and the local police killed the crocodile and recovered what was left of the young man's body.

Another hot-spot is Sesheke on the Zambezi. It was once the home of Sepopo, a king of Barotse, who had the unfortunate habit of executing people for witchcraft and other trumped-up charges by throwing them to the crocodiles. The local population of saurians understandably acquired a taste for human flesh and the crocodiles continued to take people long after Sepopo had himself been assassinated.

At one time, small boats were attacked by crocodiles in the River Gambia. This was once a flying boat refuelling stop, and people being ferried to and from the shore were sometimes confronted by unruly reptiles. It seems the aircraft engines resembled the bellows of a bull crocodile, and so a flotilla of aggrieved males would set out to confront the imaginary foe. What they found was not a rival bull but a small boat filled with tasty morsels, fresh from Europe.

There are many tales of people in boats having narrow escapes. Mary Kingsley, writing in her *Travels in West Africa*, tells of an occasion when she was trapped in a swampy lagoon and an 8ft (2.4m) long crocodile rose out of the water and placed its forequarters on the end of her canoe. Terrified, she moved to the other end of the vessel to balance the weight, and whacked the crocodile on the snout. It made off.

Crocodile expert Tony Pooley, worked as a ranger in the Natal Parks Board in South Africa, and one of the events that greeted him when he first took up his post was a crocodile attack. In December 1957, a young boy was paddling at the edge of False Bay, Zululand, South Africa, and was seized by a large crocodile. Local people reacted sharply and called for the extermination of all the crocs in the area. Crocodile attacks were to become almost an everyday event for Pooley, like the time he was inspecting for signs of poachers in the Mkonjane area of the Ndumu Game Reserve when he chanced upon a local missionary heading to a nearby village by Landrover. He jumped aboard and on reaching the settlement discovered the reason for the visit. One of a group of women walking along a hippo pathway through the reeds at the edge of a river was attacked by a crocodile. The animal tried to pull her into deep water, but her husband raced to the rescue brandishing a panga and stick. Flailing wildly at the water, he hit the crocodile with the panga and it let go, but not before it had badly damaged the woman's legs. Her right lower leg was broken in three places and splinters of bone pushed through the ripped flesh. The lower part of her left leg was perforated with deep triangular lacerations. Attended by the missionary and then shipped off to hospital, the woman survived. Chastened by the experience, Pooley then did a little research, and it transpired that in the recent past in the Kruger National Park alone, there had been three or four reported fatalities due to crocodiles nearly every summer. His study gave credibility to estimates of the annual toll throughout Africa.

In South Africa, the attacks that have occurred have taken place mainly at the same time of year, from November to April. This is the time when the air temperature is high and crocodiles are active: the rivers are swollen with cloudy, brown, flood waters in which crocodiles are virtually invisible. Small streams, normally without crocodiles, are suddenly accessible to them. Unknown to the local villagers, they move in from the larger rivers and are ready for any and every opportunity to feed.

There was, for example, the young woman who lived beside the Umsinduzi stream. Although no more than a trickle in winter, the stream swells with the summer floodwaters that flow down from the mountains nearby. On one day during the flood, the woman decided to visit her mother on the other side of the stream. She strapped her baby to her back and placed a bundle of clothes on her head, but instead of walking upstream to the usual crossing place where logs had been placed across the water, she decided to wade across the few yards to the other side. Halfway, in waist-deep water, the crocodile struck. It leapt from behind and in trying to bite her head, grabbed the bundle of clothes instead. She was knocked over, but tried to scramble to the bank. The crocodile meanwhile had discarded the inedible garments and launched a second attack. It grabbed her arm and tried to drag her into the water. She reached for a root with her free hand and prevented the animal from pulling her back. Her screams brought men running to the bank, and they threw stones, beat the animal's head with sticks, and stabbed it with an assegai until it released the woman. Her right arm was broken in two places, and the teeth had ripped muscles and skin right off the bone. Her left shoulder and chest were slashed deeply by the crocodile's claws. She was in hospital for three months. Apart from a thorough drenching, however, the baby on her back was unhurt.

Curiously, Pooley discovered that the local people had a peculiarly fatalistic attitude to crocodile attacks. Small children would swim in the shallow stretches of rivers and their parents would wade across deeper sections rather than pay a few pence to cross by boat. When attacks occurred and children were seized, the lucky survivors would be swimming at the attack site the following day. The crocodile attack was considered a providential act, and any person with a clear conscience could enter a crocodile-infested river with impunity.

Elsewhere in Africa, the same is often true. Man-eating lions and leopards cause widespread panic, and the authorities are called in to exterminate them, but a man-eating crocodile is greeted with indifference. The crocodile is after all a daily hazard for many rural Africans. Mainly a fish-eating reptile, it tends to take larger mammalian prey when it grows to a large size, say, 10–16ft (3–5m) long. The smaller crocs take small mammals and just about anything else in the water, including already dead carcasses. They are also relatively shy, choosing to avoid people. Only the big ones see us as an easy meal.

Colonel Patterson (of the man-eaters of Tsavo fame), witnessed this matter-of-fact attitude to crocodile attacks. He was walking alongside the Tsavo River with a group of Wa Kamba people when one of them went to the water's edge to fill his calabash. A crocodile rose up, grabbed the man and dragged him under. After the shouting had died down, the man's companions simply took up his bow and quiver of poisoned arrows and his stock of meat lying on the river bank and walked on as if nothing had happened.

Not long ago, this leisurely attitude to dangerous animals on the doorstep was demonstrated admirably by a baptism ceremony attended by film-maker James Barclay. He was filming missionary work in Ethiopia and on a recce for new locations when he was invited to the baptism. Throughout the ceremony, guards with rifles were posted on high ground and they took pot-shots at the crocodiles as they moved in towards the ring of worshippers in the river.

In one way, considering the number of people that are at the river's edge every day throughout Africa, the number of attacks each year is relatively low. In Tony Pooley's Ingwavuma-Ndumu district alone a population of about 80,000 people constantly visited the river. The river was and still is the centre of the community. It is where the local human

A Nile crocodile lunges at a wildebeest calf © Anup Shah, BBC Natural History Unit

Right: **American alligator underwater** © Peter Scoones, BBC Natural History Unit

Below: **Caiman with a piranha** © Ross Couper-Johnson, BBC Natural History Unit

above, left: **Zambezi (bull) shark responsible for most shark attacks in South Africa**
© Jeff Rotman, BBC Natural History Unit

below, left: **A tiger shark demonstrating aggressive behaviour** © Jeff Rotman, BBC Natural History Unit

above: **A killer whale and young**
© Jeff Foot, BBC Natural History Unit

right: **Close-up of a piranha**
© Staffan Widstrand, BBC Natural History Unit

Above: **Close-up of a common vampire bat** © Michael and Patricia Fogden, BBC Natural History Unit

Right: **Army ants on the move**
© Premaphotos, BBC Natural History Unit

population bathe, wash clothes, draw water, attend fishing nets and traps, harvest water-lily bulbs or simply ford rivers and streams.

In recent years, the number of crocodile attacks on people have been far fewer, although they still occur from time to time. In 1982, for example, a shortage of fish in Lake Mweru Wantipa on Zambia's northern border was thought to have been the reason the local crocodile population took to hijacking people. By July of that year, according to newspaper reports, they had been responsible for eating about 30 people a month.

In January 1983, an eight-year-old boy was taken by a crocodile while swimming with friends near a river dam in Natal province. The body was hidden for future consumption, but it was found later and the crocodile killed.

In 1983 and 1984, eight people were killed by crocodiles in Lake Rukwa in the western part of Tanzania. The following year, the authorities arranged for a cull of 500 crocodiles, partly to reduce the attacks on people and partly to revitalise the local fishing industry.

In November 1985, a ten-year-old girl was snatched from the banks of the Tana River in south-east Kenya, while her mother was fetching water. The mother saw the attack and ran to grab the child's leg. The crocodile had hold of her buttocks and the two had a terrifying tug-of-war, until a second woman jumped into the water and jabbed her fingers into the crocodile's eyes. The animal let go and the girl survived after a spell in hospital.

In May 1987, a 15-year-old boy was swimming in the Zambezi River in South Africa when he was grabbed by a Nile crocodile. A 20-year-old Coldstream Guards officer spotted the attack and together with the boy's father and a 20-year-old British student, he went to the rescue. The croc dragged the boy below the surface, but the guardsman grabbed and wrestled with it. He was up to his neck in water, when the boy's father thrust his arm down the creature's throat. The guardsman was able to drag the boy free. The father lost his arm up to the elbow, and the student had the bones in his forearms shattered. The boy escaped with severe bites to the arms and buttocks.

For their parts in the rescue, the guardsman was awarded the Royal Humane Society's silver medal for courage, and the boy's father and the student received bronze medals.

During 1994, the Tana River featured once again. This time a rogue crocodile terrorised residents of the village of Bakuyu in the Garissa District of Kenya. It became a problem after the Malkadaka Bridge was washed away and people had to wade across the river in order to visit friends and relatives. The croc – estimated to be about 9ft (2.7m) long – killed eight people, and took 30 goats. Local counsellors demanded it be killed. In February 1995, wardens from the Kenya Wildlife Service obliged, although it took 11 bullets to finish off the animal.

In September the same year, the residents of Yaounde in Cameroon dealt with their local man-eaters in an altogether more colourful and cruel way. Two crocodiles – a male and a female – were blamed for road accidents on the bridge under which one of them lived, and the other was held responsible for the disappearance of several children. The pair were captured by witchdoctors, dressed in human clothes (the male was even given a beard and the female was adorned with red hair and red nail polish), and then burned alive.

In February 1996, the 'Running the Nile Kayak Expedition' had to abandon its plans when the team's leaders were informed by the Uganda National Parks warden that they would be in severe danger from crocodile attacks if they attempted to travel on a particular part of the river between the Murchison and Karuma falls. A rafting expedition – the Isabindi White Water Rafters from South Africa – had tried the same stretch about six months earlier and, within 3 miles (5km) of their starting point, it had been attacked by what was described as a 'large' crocodile. Nobody was hurt, but as the raft drifted downstream the crocodile followed and attacked it again. The raft was sunk, and several people were injured but nobody was killed. The warden went on to report that there had been six croc-attack incidents during the previous year including an attack on rangers just a week before the canoeists arrived. On this occasion, the rangers escaped with their lives but not before they had to wrestle one of their colleagues from the jaws of an enormous Nile crocodile. He survived, but the crocodile managed to take a sizeable chunk of flesh from his thigh. The crocodiles, it seems, wait for dead animals that are often washed down the river, and will investigate anything that seems remotely edible.

A year after the canoeing incident, southern Somalia became the focus of world attention when unusually heavy rains caused serious flooding from October 1997 until January 1998. If the rising water, diseases, famine, and drowned livestock and destroyed crops were not enough, the local population also had to contend with attacks by the crocodiles that followed the flood waters. Just as in the rest of Africa south of the Sahara, crocodile attacks were not uncommon in Somalia. In 1994, a relief worker heard of a 25-year-old man being mauled by a Nile crocodile beside the lower Juba River. Crocodiles remained numerous despite the civil war and inter-ethnic clan warfare that resulted in a complete breakdown of law and order.

During 1994 and 1995, the authorities in Tanzania conducted a survey of crocodiles and crocodile attacks in rural areas. Most villages are close to rivers and lakes inhabited by crocodiles. Rangers covered about 620 miles (1,000km) of selected rivers, swamps and lake shores and discovered that protected areas in particular contained high densities of animals, including a 'preponderance' (according to Shlomi Ranot of Clal Crocodile Farms International Inc.) of specimens up to and over 20ft (6.1m) long.

One surprise the team uncovered was that the number of attacks on people recorded in previous years had been underestimated. One village alone – Mpanga on the upper Kilobero River, where 65 families live – lost 11 people (five of them children) to crocodiles. The survey team found similarly disturbing statistics from other villages, making Tanzania one of the top countries for fatalities from crocodile attacks. A management programme followed the survey, including the setting up of crocodile ranches, and the authorisation of hunting outside protected areas of crocodiles over 8ft (2.4m) long. The skins could be sold.

A new crocodile hot-spot is Lake Nasser, the body of water behind the Aswan Dam in Egypt. Although Nile crocodiles were known to have occupied the Nile delta and lower reaches of the river during Roman times, by the 1800s the population was completely eliminated. When the dam was built, scientists speculated that they would return, the mature lake being an ideal crocodile habitat. In 1997, evidence was beginning to accumulate that indicated this to be the case. Baby crocodiles, both living and stuffed, were being offered in the bazaar at Luxor, but more significantly, people were attacked by crocodiles. During that year there was only one fatality, but it was enough for the local authorities to introduce a 'crocodile nuisance control programme'.

Zambia has a very simple nuisance control policy: it has authorised local hunters to eliminate them. Zambia's seven major rivers – Zambezi, Luangwa, Kafue, Luapula, Chambeshi, Lusupa and Kabompo – harbour an estimated 76,440 crocodiles, according to Tourism Minister Amusa Mwanamwambwa – and crocodiles and tourism do not mix. Lives have been lost in crocodile attacks and it is bad for business.

Nile crocodiles are also found on the island of Madagascar. Here they inhabit major rivers, and are also found during the drought in cave systems, such as those at Ankarana. Just like their African counterparts, local people come down to the water to bathe, collect water, wash clothes, fish or simply to wade across. Crocodiles catch them mainly by the arms or legs, and drag them into the river where they drown before being dismembered. Most attacks take place during the dry season from November to March when rivers are at their lowest, and between 4 p.m. and 6 p.m. in the late afternoon, when women are collecting water and others bathe on their way home after working in the fields. Crocodiles are also invading small pools, swamps, and rice paddies. Between June and December 1994, 11 people were killed close to the Manimbozo and Maratony rivers, including two who completely disappeared. The number of deaths from crocodile attacks throughout Madagascar have averaged about 1,000 a year (the same as the whole of Africa south of the Sahara), according to official government figures but these are thought to be an underestimation. How the Nile crocodile made it originally to Madagascar is debatable, although it could have reached the island by crossing the Mozambique Channel. Nile crocodiles are found sometimes in estuaries and coastal waters, so a sea crossing cannot be ruled out.

The true sea-going crocodile, however, is the 'salty', as it is known affectionately down-under. The estuarine or saltwater crocodile (*Crocodylus porosus*) is so at home in salt water it can be found far out at sea. Indeed, a specimen was once spotted 30 miles (48km) north of New Zealand's North Cape, another at the Cocos (Keeling) Islands about 600 miles (970km) from the nearest landfall, and on 18 September 1994 swimmers in the ocean at Nauru, about 620 miles (1,000km) north-east of the nearest known crocodile population in the Solomon Islands, spotted a saltwater croc alongside them in the ocean. A youngster, it was caught and taken to the local police station.

More usually the saltwater crocodile is to be found in inshore waters and estuaries from Cochin on the west coast of India, in the Sundarbans, across the Indian Ocean to New Guinea, south to Queensland, Australia, and further to the east around the islands of the south-west Pacific as far as Fiji.

This species is the largest and potentially the most dangerous of all the crocodiles. In fact, it is the planet's largest living reptile. It is estimated to grow up to 30ft (9m) long, and may weigh over 2 tonnes, although giants of this ilk are unlikely to have survived the hunter's gun. During the early part of the 20th century, however, a monster is reputed to have lived in the Segama River in northern Borneo. The local villagers – the Seluke people – claimed it to be over 200 years old. They were in awe of the beast and threw silver coins into the river in order to placate this 'father of the devil' as it was known. An impression its body had made on a sandbank was measured, and the beast was found to be 32ft 10in (10m) long.

The largest accurately measured specimen was shot in 1957 by Mrs Kris Pawlowski, wife of crocodile expert Ron Pawlowski. The creature was on the MacArthur Bank in the Norman River, south-east of the Gulf of Carpentaria in northern Australia. It was 28ft 4in (8.64m) long.

The most famous salty in Australia was one dubbed 'Sweetheart' in the late 1970s. This 18ft (5.5m) male, weighing 1,800lb (816kg), lived in the Sweet's Lookout Billabong on the

Finniss River, Northern Territory. It acquired notoriety for its inclination to grab outboard motors. It destroyed more than 20 engines on small craft, and even ripped off the sides of boats, sinking one or two. Though Sweetheart never attacked a person, local fishermen were understandably concerned and threatened to shoot the beast. In July 1979, however, state wildlife officials tried to relocate the crocodile in a wilderness area but it became trapped under a log while still woozy from the anaesthetic darting and drowned. Its stuffed body can be seen today at Darwin Museum.

While Sweetheart was a recognised destroyer of boat engines, other saltwater crocodiles have been undisputed man-eaters, and in the parts of the world where salties are common, local people treat them with great respect; yet they still get caught. One of the earlier reports dates from 1823 when Frenchman Paul de la Gironere, who spent many years in the Philippines, came across a 20ft (6.1m) long man-eater at Jala Jala, near Lake Taal on the island of Luzon. The crocodile's last human meal had been a local shepherd who had tried to wade across the river. De la Gironere and an American friend set out to capture the beast. After several days, they eventually trapped it and, after a six-hour struggle, killed it by severing its spinal chord with a sharp lance. It was so heavy, 40 hunters only just managed to move it. In its stomach were seven or eight pieces of a horse and a pile of pebbles.

More recently, the Philippine island of Siargao was under siege from a single crocodile that managed to grab at least nine people. And, to the east of the Philippines, in the south-west Pacific, a fisherman on the island of Palau was killed by a saltwater crocodile early one morning in 1965. When his body was recovered, the man had one arm missing and part of another, and his liver and a lung had been taken.

Further to the south, Sarawak has been host to some giant man-eaters. The Lapar River, for example, was the site of a number of attacks, six of them fatal, between 1975 and 1984. And fishermen on the Batang Lupur River have been attacked by crocodiles. The main culprit was an enormous individual, estimated to be about 25ft (7.6m) long, and known to the locals as the 'king of crocodiles'. It was clearly identified as the beast responsible for many incidents for it was easily recognised by the white patch on its back. Curiously, though, the attacks were seasonal, between May and September, and always in a particular part of the river, between Lingga and Bijat. When researchers began to analyse the historical records and match this with the biology of the saltwater crocodile, they realised that the local prawn-fishing season coincided with the main breeding period of the crocodiles. The crocodiles were aggressively territorial during this period, and the fishermen were regularly coming into conflict with the local population. Down the years, many people lost their lives.

The next major island to the east of Sarawak and Borneo is Celebes, home not only to giant snakes but also to saltwater crocodiles. In December 1975, a tourist boat sank in the Malili River in Celebes and passengers and crew were tipped into the water. Crocodiles attacked and ate 40 of them.

Further to the east again is the island of New Guinea, with Irian Jaya occupying the western half and Papua New Guinea in the east. Here, crocodile fanatic George Craig recalls the story of a young man and his wife travelling along a river in eastern Papua New Guinea. The wife was seated at the back and was taken by a large crocodile. The man dived on the animal and tried in vain to free his wife. As they all went under, he jabbed his thumbs into the crocodile's eyes and it released the woman. The two made it safely to the river bank.

Evelyn Cheesman draws attention to another gruesome event in her book *Six-Legged Snakes in New Guinea*. The year was 1939, and she chanced upon a group of villagers who had trapped and killed a 16ft (4.9m) crocodile that had been hiding under an overhanging river bank. It was suspected of being a man-eater. When its stomach was opened, inside was the partly digested torso of a man who had vanished from the water's edge the previous day. His legs were missing.

During the Second World War, American entomologist P. J. Darlington was collecting mosquito larvae in New Guinea when he was attacked by a crocodile. He tried to escape up the bank but slipped and was grabbed by the arms. He fought long and hard, and eventually the beast released him and he lived to tell the tale.

Since the war, saltwater crocodiles have continued to take the occasional human, although in some places 'occasional' has been replaced by 'frequent'. In the 1960s, for example, a village in northern Irian Jaya was reported to have been terrorised by a solitary crocodile. More than 60 people were killed or injured.

In Papua New Guinea, the Barema River of west New Britain became the focus of press attention when a 13-year-old boy, who was swimming with his family, became the 13th victim of an unusually large saltwater crocodile estimated to be 33ft (10m) long and, according to eyewitnesses, as wide as a 44-gallon oil drum.

Many of today's man-eating reports, however, emanate from Australia where people involved in fishing and water sports are coming into conflict increasingly with salties. In April 1981, a 23-year-old man and a 12-year-old girl were on an airboat excursion around the swamps near Channel Point, about 112 miles (180km) to the south-west of Darwin, Australia. The young man dropped his pistol in shallow water and as he bent down to retrieve it, he was caught by a 13ft (4m) long saltwater crocodile. In the struggle the crocodile grabbed the man's left arm, then his right thigh and finally his right buttock, continually spinning and trying to drag him into deep water. Bravely, the young girl plunged into the water and held on to the man's right arm. By this time his head was below the water, but the girl would not let go. The crocodile eventually tired and released the man, and the girl was able to help him to the top of the bank. The crocodile floated offshore and watched for another opportunity.

Fortunately, the girl had already learned to drive and so, leaving the young man in a safe place, she ran across country, where other crocodiles could have been lying in wait, to fetch a truck. After applying antiseptic powder to his wounds, she drove him towards Darwin, meeting another vehicle that they had been able to summon by phone from a homestead 20 miles (32km) away. The man recovered in hospital, and the girl was awarded the Royal Humane Society's Clarke Gold Medal for what was described at the time as 'the country's most outstanding incident of bravery'.

Saltwater crocodiles can turn up in the most unexpected places. During the wet season they even wander into Cairns, a sizeable town on the Queensland coast. After heavy rains in May 1983 a crocodile reared up from the gutter and grabbed a 19-year-old man by the leg. A passing taxi driver helped him to fight the creature off.

Heavy rains in February 1998, saw crocodiles entering the suburbs again, this time Westcourt, a suburb of Brisbane. A 15-year-old schoolgirl was grabbed from beside a small creek and dragged into the water as the creature went into a death roll. Her grandmother raced after her and, as the crocodile surfaced momentarily and the girl grabbed an overhanging branch, she screamed out and snatched the teenager from its jaws. The girl's legs were shredded and an ankle broken but medical teams at the local hospital worked hard to save them. Crocodile experts thought she was lucky to be alive. The attack followed a similar one on a drunken swimmer in the same area in December 1997, and was followed by the sighting of a crocodile in a storm-water drain in the same area in April 1998. In fact, things became so dangerous and attacks were so frequent that, in April 1998, plans were made by the Queensland government to remove and relocate crocodiles from waterways around the city.

Some of the worst crocodile incidents in recent years, however, have been the result of unwitting tourists behaving in what the locals would consider foolish ways in crocodile country. In February 1985, for example, a woman was on a walking holiday alone in the Kakadu National Park, in northern Australia, when she was chased and attacked by a 12ft (3.7m) crocodile. She was badly hurt, and had to crawl for five hours a distance of more than a mile before she was found.

Towards the end of the year, in December 1985, a 43-year-old woman was wading in shallow water at Barratt Creek, about 950 miles (1,530km) north of Brisbane, when she was seized by a crocodile. The culprit, a 15ft (4.6m) salty, was caught a few weeks later and inside its stomach were what were thought to be her last remains – fingernails, toenails and a few bones.

In February 1986, the dismembered body of a 26-year-old fisherwoman was found near a 17ft (5.2m) crocodile on the swampy bank of the Staaten River in northern Queensland. The woman had vanished when she and a male companion had swum from their dinghy, that had been stranded on a sandbank, to their fishing boat anchored a short distance away. The man hauled himself aboard and turned to help the woman when the large crocodile grabbed her and swam away. The crocodile was trapped and killed, as were many more in a spate of killings to rid the area – a tourist spot – of all its large crocodiles.

In the following year, during the southern summer of 1986–7, there was a spate of crocodile attacks in northern Australia. Many involved visitors who came to the region after the success of the feature film *Crocodile Dundee*. In September 1986, the severed legs of a man were all that remained after a crocodile had taken him while he slept in a camp outside a fishing village in the Northern Territory. The rest of his body was found partly digested in the stomach of a 15ft (4.6m) long salty which was trapped with nets and harpooned a few days later.

Early in March 1987, a party of American tourists visiting the Northern Territories watched in horror as a crocodile attacked a fisherman wading across the East Alligator River. That same month, a 24-year-old American model from Colorado on a cruise along the coasts of northern Australia was grabbed by a crocodile, estimated to be about 12ft (3.7m) long. The young tourist had been basking in a rubber dinghy and tried to swim back to the cruise launch moored in the mouth of the Prince Regent River to the north-east of Broome, Western Australia. Eyewitnesses reveal that they saw the crocodile surface near the woman and, following a commotion, both disappeared. Her body was recovered some while later, and when it was being carried back to shore another crocodile leapt from the water and tried to snatch it from the boat.

Later in the year, the foot of a fisherman who had mysteriously disappeared a few weeks earlier was found inside a saltwater crocodile that had been shot by hunters. At

Weipa, in northern Queensland, a 32-year-old man was taken by a salty as he went to cool off in a shallow creek. In fact, a total of seven people were killed during a 15-month period, and in view of the impact the attacks would have on tourism the Queensland authorities threatened to capture every crocodile over 4ft (1.2m) long between Cooktown and Rockhampton and transport them to zoos and wildlife parks. They also suggested stiff fines for those people that put themselves at risk.

One of the most horrific incidents involving saltwater crocodiles, however, occurred in February 1944 when allied soldiers cut off a large contingent of Japanese troops on Ramree island, close to the coast of Burma in the Bay of Bengal. A corridor of vegetation runs the 18 miles (29km) from the island to the mainland, but the ground here is no more than crocodile-infested swamp. In order to escape from the British army advancing across the island, the Japanese made for the swamp, hoping to be picked up by their own ships. The Royal Navy blockaded the island, however, and so they were trapped. The British also used artillery and mortars to contain the Japanese troops. As the tide rose, the noise and confusion drove the local crocodile population into the areas of clear water within the swamp, and there they lay like lifeless floating logs. When the tide went out again, some Japanese had escaped but it had left others trapped in the swamp. During the night, the crocodiles moved in. The thrashing and screams of their victims could be heard from the British lines and nearby ships. Of an estimated 1,000 troops that went in, only 20 are known to have come out. Some authorities believe that many, indeed, did escape to the mainland, but many were also killed by crocodiles. Witness to the entire episode was naturalist Bruce Wright, and he recalls the event in *Wildlife Sketches, Near and Far.*

In the most westerly part of its range, people put themselves at risk just by trying to make a living. On the coasts of India, the saltwater crocodile is yet another man-eater with which its population must contend. During the 1970s, for example, the Indian state of Orissa recorded four fatal attacks by saltwater crocodiles in ten years. Two attacks, that occurred three years

165

apart, were thought to have been by the same crocodile. Interestingly, this individual could have taken people on a regular basis, but it failed to do so. This indicates, perhaps, that saltwater crocodiles do not target people specifically, and do not acquire a taste for human flesh.

In the adjacent state of Bengal, the crocodile has joined the tiger in terrorising the fishermen and woodcutters of the Sundarbans. Elsewhere in India other species of crocodile have acquired evil reputations.

The gavial (*Gavialis gangeticus*) of the Indian subcontinent grows to enormous size, up to 20ft (6m) long, but as it is mainly a fish-eater, it is not thought to be harmful to people. Nevertheless, people have been attacked. Around 1950, a fisherman was attacked by a gavial at Talchar on the Brahmani River, and in 1974 a fisherman from the village of Naraj on the Mahanadi River dived into the water to release a trapped net and was grabbed around the chest. In the same year a visitor to Naraj was bitten on the ankle. In November 1979, a 55-year-old man was doing his washing up in the Mahanadi River, in the Satkoshia Gorge Sanctuary, when his arm was gripped in the long, slender jaws of a gavial estimated to be about 13ft (4m) long. The man somehow ended up in the water, having lost consciousness. As he was being hauled out by the hair by his son in a canoe, the gavial attacked again, this time biting the man's thigh. As in all the other recorded gavial attacks, the animal was not thought to be feeding. It was probably defending its patch.

The mugger or swamp crocodile (*Crocodylus palustris*), also living on the subcontinent, is another crocodilian that has been labelled 'man-eater', but the evidence for it being a regular one is slim. It is regarded as a less aggressive animal than both the Nile and saltwater crocodiles, but it is a sizeable animal, growing up to 16ft (5m) long, and so must be considered a potential danger to people. It acquired a bad reputation from its supposed habit of scavenging on the human corpses that float downstream from Hindi cremation sites on the river bank. Bracelets, earrings and anklets found in the stomachs of muggers are witness more to this scavenging habit than to predation. However, local myth has it that, having eaten the flesh of human dead, the mugger takes to catching living people.

There have been a few well-documented attacks on humans. In 1960 a mugger was reported to have attacked and consumed a young boy, and in 1965 a beggar sleeping on a footpath close to a reservoir was taken and eaten. Both attacks took place in the Indian state of Gujarat where there are many reservoirs and ponds containing crocodiles. Here many people and a few crocodiles are in close proximity and so the risk of attack is always present. Surprisingly, though, there are only a few attacks reported. In May 1991, for example, a man was attacked by a mugger at the Rudra Mata Dam near Bhuj in Kachchh (Kutch), north-west India. For many months, crocodiles had been taking livestock, including goats, sheep, and chickens but, gradually it is thought they lost their fear of humans and turned to man-eating. Earlier the same month, a mugger crocodile – a male 11ft (3.36m) long and weighing 463lb (210kg) – killed a 10-year-old boy near the village of Dundelav in the Vadodara District. It was captured and taken to the Sayaji Buag Zoo.

In the New World, the American crocodile (*Crocodylus acutus*) is about the same size as the mugger and it too has been implicated in attacks on people in countries bordering the Caribbean and the Gulf of Mexico, particularly in Mexico itself. In August 1993, an eight-year-old boy was grabbed by a 9.8ft (3m) long American crocodile. The boy was playing in a small river about 5ft (1.5m) wide and just 1.6ft (0.5m) deep about 66ft (20m) from his house near Puerto Vallarta, Jalisco, on Mexico's west coast. The animal held the boy's head and left arm in its jaws and was trying to drag him away. Several passers-by threw stones and branches, and the crocodile eventually let go. The boy was scared but not hurt. Some time later, enquiries by two crocodile researchers revealed that the crocodile was probably a captive animal that had escaped about three months previously, and it was therefore unafraid of humans.

During 1996, there were reports of seven attacks by American crocodiles on people in the state of Quintana Roo on the eastern side of the Yucatan – six in the Nichupte Lagoon in

Cancun, and one in the Sian Ka'an Biosphere Reserve south of Cancun. All were either fishing or spearfishing and none of the attacks was fatal, although several came away with injuries including one person losing part of his foot.

The attacks were thought to be attributed to just three enterprising individuals, from the lagoon's large crocodile population, which were fed scraps of food by restaurant owners at one end of the lagoon. The practice had become a tourists' attraction. In the crocodile's eyes, therefore, people are synonymous with food. Couple this with the fact that at least four of the attacks took place during the crocodile's breeding season, when the animals are more aggressive anyway, and you have a recipe for a disaster. This is precisely what happened in 1996. Two attacks occurred within a day of each other when snorkelers, who were spearfishing illegally, swam in a 66ft (20m) wide canal lined with red mangroves. Another two attacks on spearfishermen took place under a bridge at the opposite end of the lagoon, just a week apart. And attacks five and six were launched on people fishing at roughly the same spot at the water's edge, also a week apart. One lost a foot in the attack and another was grabbed by the head. Attack seven at Sian Ka'an was on a woman spear-fishing with her partner. She was also grabbed by the head. Nobody was killed.

This recent spate of crocodile attacks follows others in previous years. The other species thought to be responsible was Morelet's crocodile (*Crocodylus moreletii*), another Central American species. It grows to about 9.75ft (3m) long and inhabits swampy areas of freshwater where it more usually feeds on freshwater turtles and molluscs. In 1992, however, one Morelet's crocodile took a fancy to something a little more interesting. A fisherman was attacked while wading in shallow water at the Laguna Muyil within the Sian Ka'an Biosphere Reserve, and two years later his brother was grabbed first by the head and then by the legs in roughly the same spot. It is thought the same crocodile was responsible for both attacks. Then, in 1995, a female crocodile grabbed the hand and thigh of a fisherman standing in water up to his waist. He had grabbed a Morelet's hatchling and was most probably attacked by the mother. The incident took place to the south of the El Eden Ecological Reserve. So far, there have been no fatalities.

The aggressive Cuban crocodile (*Crocodylus rhombifer*) rarely grows to more than 6ft (2m) long, and is not thought to be a threat to adult humans. However, its habit of leaping clear of the water when pursuing prey could mean that children playing on river banks are vulnerable.

The Orinoco crocodile (*Crocodylus intermedius*), which lives in the Orinoco and Meta rivers of South America, may grow to 22ft (6.7m) in length and has been identified as a potential man-eater but there are few cases to suggest the habit is widespread.

The black caiman or jacare-acu (*Melanosochus niger*) is also a large animal, with lengths up to 19.7ft (6m) recorded. Attacks are rare, but local people are careful when going down to the water's edge in areas where these reptiles are present. Henry Walter Bates, who travelled extensively in Brazil and wrote the classic *The Naturalist on the Amazon* in 1892, reported only one authenticated case of an attack on a human. The local justice of the peace at Caicara watched as a drunken man started to wade into the muddy river where a large black caiman was known to live. He called to the man to return to the shore, but he was ignored. Seconds later the monster rose up from the water's surface and seized the man around the waist. He was pulled below and never seen again.

More recently, in February 1997, a researcher was attacked by a black caiman. He was one of four people in a 23ft (7m) long canoe, and they were out collecting caiman hatchlings at Imuya Lagoon in the Rio Napa region of eastern Ecuador. The man was standing in the boat with bent knees and turning to watch one of his colleagues when the female caiman launched at him. She rose about 3ft (1m) out of the water, grabbed at his raincoat and left buttock, and pulled him backwards into the water. He was dragged down about 6ft (2m) before being released. As he bobbed to the surface and swam frantically for the canoe, the caiman reappeared. It was facing away from the boat, and when it turned, the sweep of her tail pushed the man back under the water. When it came up again, one of the other scientists clouted it on the head with a small paddle. The man scrambled hastily back into the canoe. He had a puncture wound and a large bruise the size and shape of a caiman's jaws. Clearly, the female was aggressively protecting her offspring, but it did give some credibility to the suggestion that the black caiman could be a potential man-eater.

The acknowledged man-eater in the New World, however, is the adult American alligator (*Alligator mississipiensis*), an animal that has eaten more Americans than have been killed by sharks during the past 50 years. Males may grow to 20ft (6m) long, although all large alligators of either sex are quite capable of pulling a person into the water. It lives in the southern states of North America where one of the first references to it attacking humans was in 1685. Dumesnil, a servant of René-Robert Cavelier – the Sieur de la Salle – was swimming across the Colorado River in Texas, when he was seized by an alligator. Another incident on 10 August 1734, which was recorded in a Louisiana courthouse, involved a blacksmith who was swimming in the Red River and was killed by alligators.

The alligator's evil reputation, however, became more widespread some time later, at the beginning of the 19th century. This was the time when the south-eastern states were being colonised and settled. Its victims were Negro slaves. It was they and not their white masters who went to the rivers and streams to draw water or wash clothes, and while they went about their chores they were snatched from the water's edge. The Mississippi State Plant Board, for example, records a black child being grabbed from right beside his mother at Alligator Lake, Bolivar County. Could it be that hundreds of black slaves suffered the same fate, their disappearances unrecorded?

In 1836, refugees from the Mexican wars were exposed to the same fate. In a US technical bulletin dated 1929, for example, Remington Kellog tells of the King family's horrific ordeal. They were crossing the Trinity River in Texas when their ferry ran aground. The head of the family jumped into the waist-deep water and before he could push the boat free he was taken by alligators. During the same campaign others were attacked, but a young black servant who had been sleeping outside his master's tent survived. His master, one James Kerr of Kerr County fame, took a burning branch from the campfire and thrust it into the alligator's eyes. Reluctantly it released its prey. A similar act of bravery occurred on the Brazos River, when Ben McCulloch of McCulloch County fame, gouged the eyes of an alligator and encouraged it to release a young soldier which it had just seized. The result of these kind of attacks was the wholesale slaughter of alligators, and so they became wary of people. Children were still vulnerable, though. A report from 1928, describes how a young boy was grabbed by an alligator and drowned in a pond on a golf course.

The resurgence of the American alligator, however, was not until the end of the Second World War, when a new era of nature conservation resulted in an upswing in the alligator population. It was also accompanied by an increase in the number of attacks on people. In 1948, a 9ft (2.7m) long male savaged both the arms of a woman swimming in a deep Florida spring. The woman survived, the alligator was killed. Two years later, in April 1950, a 15-year-old girl was grabbed by the foot whilst swimming in a Florida creek. Rescuers managed to pull her into a boat and she was saved.

In 1949, fishermen and tourists visiting the Okefenoke Wildlife Refuge had the unnerving feeling of being stalked by two particularly large and menacing alligators. The duo were trapped and removed from the swamp in case they should turn stalking into eating.

More incidents were reported throughout the 1950s, culminating in two horrific cases in 1957 and 1958. The first involved the disappearance of a nine-year-old boy in 1957. His mutilated body was later found in a creek in Brevard County, eastern Florida. Alligators were the prime suspects and several large specimens were shot, two with what were thought to be human remains inside. A year later, a swimmer snorkelling in a lake was gripped by the head, and was badly slashed by an alligator's teeth as he tried to reach the safety of the shore. He survived. And in 1959, Daytona Beach was the site of another horrific attack during which a youngster was pulled under and drowned.

Attacks continue to this day. Often the attack sites are places where alligators have been used to receiving free hand-outs. They have no fear of people, and anything that should enter or fall into the water is considered food. Just such a thing happened in 1973, when a teenaged girl went swimming in a lake in central Florida. She was attacked and killed by an 8ft (2.4m) alligator that was well known to local people and considered harmless. In 1975, a field biologist was working in the Oklawaha River, when an alligator estimated to be about 12ft (3.75m) long set about him. In September 1977, a 52-year-old man was savaged and killed by a 7ft (2.1m) long alligator, and in September 1978, a 14-year-old was the victim of an alligator attack. In 1984, an 11-year-old boy was swimming in the St Lucie River, Florida when a 12ft (3.7m) long alligator seized him and pulled him below the surface. Police shot at the beast and it dropped the boy but he was already dead.

In 1987, an English student from Cranbrook was water skiing on a lake near Daytona Beach. He fell from the skis into the water and was supported by his life vest about 50ft (15m) from the shore when he was grabbed on the right ankle by an 8ft (2.5m) long alligator. 'I could feel teeth,' he recalled. 'There was a lot of pressure. I could feel him pulling me right down. It was pretty weird. I dug my fingers into his snout and pulled as hard as I could, but at first it seemed useless. I thought it was all over and I thought about all the things I should have done, then suddenly it released me.' The young man started yelling, but his friends thought he was joking. 'When he started shouting " 'gator", we all started laughing,' said one of his colleagues from Edinburgh. 'We thought he was kidding, but when he kept on, we realised he was in trouble.' He suffered eight-inch-deep puncture wounds on his right foot, torn skin on the back of his heel, and bruised arms. The alligator was caught and killed.

Between 1978 and 1988, 71 people were attacked by alligators in Florida, and they turned up in the most unexpected places. Theron McBride, for example, made his living by collecting golf balls that had dropped into the man-made ponds on golf courses. One day, he was in a pond between the sixth and seventh holes of the Ponciana County

Country Club at Lake Worth, Florida, when he was seized on the flipper by an 8ft (2.4m) long alligator. The animal pulled him to the bottom, but McBride was able to twist his foot free and escape. The local wildlife officer was called and, after a struggle, he had caught not one but three alligators in the pond. At another Florida golf course a player went missing one day. His friends suspected that he had been taken by an alligator known locally as Ol' Mose. The animal had frequented the course for over 20 years without incident, but when it was trapped, killed and sliced open, the missing golfer was found inside.

The influx of people searching for the winter sun in Florida has inevitably resulted in accidents with the local wildlife, and, poisonous snakes like the water moccasin aside, a big alligator is a formidable hazard. A toddler falling into a canal or a bather swimming where he or she should not be are prime targets for such an alert beast. The incidents are few, but each one is undoubtedly frightening. In the summer of 1987, for example, a scuba diver swam outside a roped-off safe swimming area at Wakulla Springs State Park and disappeared. Some time later, a tourist boat was plying the Wakulla River and those on board were horrified to see an alligator pass by with the dead body of the diver in its jaws.

On 4 June 1988, a four-year-old girl was walking her dog along the shore of a lake at Port Charlotte when a 10.5ft (3.2m) long alligator rushed out of the water and grabbed her. The girl died. In June 1993, a ten-year-old boy who was wading in shallow water in the Loxahatchee River was knocked over by an alligator and dragged downstream. His father and several eyewitnesses attacked the animal with paddles, but the boy was already dead. And, in October the same year, a 70-year-old woman was torn apart by alligators at Lake Serenity, Sumter County, Florida. When state officials killed some of the 'gators they recovered most bits of the woman, except her right arm.

In October 1997, a 35-year-old snorkeler had a narrow escape when he was swimming at Juniper Springs during a canoe trip. The young man was in waist-deep water and had just put his head under when he was face-to-face with an alligator. He saw its head move towards him and its mouth open before the crocodile lunged and grabbed him. His head was inside the crocodile's mouth and it shook him from side to side for about 30 seconds. The man managed to punch the animal in the throat and it released him. Friends hauled him out of the water. The crocodile's teeth had punctured his chest, collapsing one lung, made deep dents in his head and neck. The alligator was later killed by trappers.

With so many people moving in to live in Florida (an estimated 800–1,000 people per day), conflicts between man and alligator inevitably are going to increase. The case of the seven-year-old boy who fell off his bicycle and ended up in a water-filled ditch in the Everglades National Park is a sad reminder that alligators and people do not make easy neighbours. The alligator mangled an arm and bit into the chest. This conflict was further illustrated by an attack on a small boat by an alligator. The owner felt his craft tip backwards. Teeth fragment and scratch marks identified the assailant to have been an alligator.

About 200 or so people are attacked each year, of which a half-dozen are fatal. The consequence is that state officials take every report of a life-threatening alligator very seriously. The 'nuisance alligator control programme' receives about 15,000 calls each year, and about 5,000 alligators are killed legally by licenced state trappers.

The longest alligator ever to have been caught (and the longest on record) was a specimen over 14ft (4.3m) long which was captured in Lake Monroe on 30 December 1997. Local residents had labelled it a 'nuisance alligator' and thought it could be dangerous to people. After its capture, a buyer could not be found for it to be kept in captivity and so it was killed.

Florida, however, does not have exclusivity in alligators. Other southern states have their fair share, and occasionally they make themselves felt. At Hilton Head, South Carolina, for example, a 69-year-old man went into his garden to restock his bird-feeder when an 8ft (2.4m) long alligator from a nearby lagoon attacked him. The man began to run but tripped and fell. The alligator grabbed his foot. The man managed to get away, but the animal was later trapped and shot.

CHAPTER 10

# DRAGONS AND SERPENTS

While crocodiles, alligators and caiman represent the largest and most powerful man-eaters amongst the reptiles, there are others. Several species of lizards and snakes, for example, grow large enough to pose a real threat to people.

The Komodo dragon (*Varanus komodo*) is the world's largest lizard, with a maximum known length of 10ft (3m) and a weight of 365lb (166kg) and the suggestion that it can grow even bigger. A Swedish naturalist, for example, found what was estimated to be a 23ft (7m) specimen on a Komodo island beach, and an American newspaper reported someone having seen one more than 14ft (4.3m) long. Today, male dragons rarely exceed 8.5ft (2.6m) and females 7.5ft (2.3m). Whatever their size, they are potentially dangerous to humans. The danger, however, from these monster lizards is not necessarily in their size and ferocity but in their mouth.

The Komodo dragon is mainly a scavenger, feeding on the decomposing carcasses of feral goats and wild deer, so its mouth tends to be home for extremely nasty bacteria. When the animal turns to hunting, it uses this frightful flora to help it catch a meal. It will lie in wait for unsuspecting prey and then race out and bite it on the leg. The wound becomes infected, and eventually the victim succumbs to blood poisoning. Researchers have identified at least four types of bacteria producing toxins for which there are no known antidotes. All the dragon has to do is wait for the prey to die. Local people, including children, have been on the list of casualties. In 1994, an Indonesian policeman was killed by dragons.

The Komodo dragon is a ravenous meat-eater. It is one of the few reptiles, apart from turtles, that cuts its prey into large pieces before swallowing them whole. In one sitting it can consume an amount of meat almost equal to its own body weight, swallowing down large 5.5lb (2.5kg) chunks every minute. A 110lb (50kg) specimen was once observed downing a 68lb (31kg) pig in just 17 minutes. An enormous gall bladder helps it digest extraordinarily large meals – the equivalent of a person scoffing over 600 hamburgers in one sitting. It bolts its food rapidly, before another dragon can come along and steal it. It has sharp claws and serrated teeth: any teeth that are knocked out are replaced.

It eats almost anything that it can either ambush or scavenge – rodents, snakes, birds, deer, feral goats and pigs, injured and weak buffalo, and even smaller members of its own kind. Hoofed and horned animals are swallowed in their entirety. In ancient times, before people brought pigs and goats to the island, it is thought these giant lizards ate pygmy elephants (*Elphas*), pig deer (*Babyrousa*), endemic wild cattle (*Anoa*) and a giant tortoise (*Geochelone*).

Today, dragons have been seen to follow a pregnant goat and wait for the new-born kid to drop before going in for the kill, sometimes taking both baby and mother. They rely on smell to seek out putrefying flesh on dead carcasses some distance away, and have also been known to dig up fresh graves in the cemetery and consume the contents.

Some of the dragons on Komodo are fed regularly at baiting sites for visitors and these individuals tend not to attack people. Tourists visiting the island can watch them feeding in relative safety. Away from the tourist areas, the picture is a quite different one.

The world authority on Komodo dragons, Dr Walter Auffenberg, now retired from the University of Florida, recalls a time in 1969 and 1973 when he was staying on Komodo and Flores islands with his family and came across 'rogue' dragons which turned out to be particularly dangerous. There was one large male, which was given the study identity number '34W', that Auffenberg is certain had killed three people in the Poreng area of Komodo. The local people nicknamed it 'ora gila' meaning 'crazy monitor'. Footprints on the beach indicated that the lizard had been stalking Auffenberg's own children, and one night it walked casually into the family's tent and shredded a bag of clothes. The family ran for safety.

Some days later, a tourist walking through 34W's home range disappeared. All that was found was a blood-stained shirt and a camera case. The man was never discovered. Baron Rudolph, an 89-year-old German-Swiss traveller, disappeared in similar circumstances. He strayed from a line of people and into the bush. Only his camera and glasses were recovered. Records indicate that six other tourists have been lost to dragons in the past 20 years.

Local villagers have great respect for the animals and try to stay out of their way. The problem is that they tend not to chase their prey (although they can move at about 9mph or 14km/h in short bursts) for they quickly overheat, but wait in ambush. Anybody walking alongside a stand of long grass or bushes in which a dragon is hiding is vulnerable to attack. One story tells of a father and two sons who were cutting wood in the forest, when a Komodo dragon raced towards them. They got up and ran, but one of the boys crashed into a low-hanging vine and the dragon grabbed him by the buttock. It tore away a large chunk of flesh and within the hour the boy bled to death.

The reptiles are also attracted to villages. A young girl on Komodo was climbing down the ladder from her house built on stilts when a dragon that had been hiding in the shadows grabbed her and killed her. On Flores, a 10-year-old boy was grabbed from the veranda of his house. His mother struggled with the dragon and forced it to release the boy, but its jaws had sunk deep and the boy was already dead.

Dragons are not restricted to Komodo and Flores. They are also found on the Indonesian islands of Rintja, Padar, Gili Mota, and Owadi Sami – where they are confined to nature reserves. But on the islands of Papua New Guinea another dragon vies with the Komodo dragon for the title of world's largest lizard. It is called the Artrellia. Lengths of up to 30ft (9m) long have been claimed for the beast, but science has yet to recognise that it exists at all. According to local people, it is an opportunistic man-eater, eating everything and everyone that should stumble into it.

In 1980, explorer John Blashford-Snell was presented with what was professed to be a young Artrellia, but it turned out to be a specimen of Salvadori's monitor (*Varanus salvadori*), the longest (but not the largest) living lizard which grows up to 15ft (4.6m) long. It is much more slender than the Komodo dragon, being mostly tail, and therefore is not as bulky.

The Artrellia may be an island myth or simply Salvadori's monitor, but a monster monitor lizard that measured 23ft (7m) and weighed up to 2,205lb (1,000kg) was a reality in Pleistocene times in Australasia. This was *Megalania* and it is now extinct . . . or is it? Could it be alive and well and living in the wilds of Papua New Guinea?

The word 'dragon' originates from the Greek word for 'seeing with keen eyes' and was used by the Greeks, according to Konrad Gesner, to describe 'snakes, especially for those snakes that are large and heavy and surpass all others in size'. Indeed, learned scholars who described 'dracones' were referring to giant python-like snakes rather than lizards. Pliny mentioned them too. He wrote that 'they lived in India and they fell down from trees upon their victims, whom they killed by encircling them in their coils'. These became better known as 'serpents', and there have been reports of monstrous ones which attack and swallow people whole.

South America has what has become the world's most notorious snake – the anaconda (*Eunectes murinus*). Although not as long as some of the Old World's constricting snakes, it still grows to an immense size and is powerful and potentially dangerous.

Anacondas spend all their time with their enormous bulk supported by the water in lakes and slow-flowing rivers in tropical South America. The invading Spaniards wrote about 60–80ft (18–24m) long anacondas which they called 'matatora', meaning killer of bulls.

There are many giant anaconda stories, but most must be taken with a good deal of scepticism. In 1948, for example, a 130ft (40m) anaconda was supposed to have been captured alive by local Indians beside the River Amazon. It was towed to Manaos where some lunatic, in true frontier style, shot it. Later that year a 115ft (35m) snake was killed supposedly at Abunda, and in 1954 a 118ft (36m) specimen was shot allegedly at Amapa. Some time previous to this, in 1907, Colonel P. H. Fawcett shot an anaconda, which he claimed was 62ft (18.9m) long, on the Rio Abunda.

The largest accurately measured anaconda was one shot on the Colombia–Venezuela border in the 1940s. Members of an oil exploration team said they spotted a snake in the Upper Orinoco River. They shot at it, dragged it from the water, and measured it. It was 37ft 6 ins (11.43m) long. They left the snake on the river bank while they continued with their work, but when they returned to remove the skin they discovered the snake was gone. It was not dead at all; it had been only stunned by the bullets.

The discovery of such large animals in South American rivers and swamps meant that the inevitable man-eating stories followed, but substantiated incidents are few and far between. One of the more recent accounts was in July 1979. A fishermen was trying to untangle his net in a lake near Racing in north-west Brazil when he was grabbed by a huge snake, probably an anaconda, said to have a body 2ft (60 cm) wide. The remains of his body, showing a crushed torso and most of the flesh ripped away, was later washed ashore.

Of the other large constricting snakes, a size of 130ft (39.6m) was claimed for an African rock python (*Python sebae*) killed by Bwanba tribesmen in the Semliki Valley, Central Africa, in 1932; a 36ft (11m) reticulated python (*Python reticulatus*) is said to have been captured in Celebes as recently as 1971; and a 25ft (7.6m) amethystine python (*Liasis amethystinus*) was killed near Cairns, Queensland, Australia.

Clearly, even believably large snakes can be quite formidable and dangerous. Anacondas and rock pythons have been found with caiman and crocodiles in their gut. There was, for example, the African rock python shot by K. H. Kroff in Northern Rhodesia in 1958. Inside the 23ft (7m) long snake was a 5ft (1.5m) long Nile crocodile. Others have been found containing whole pigs and peccaries. The largest intact item of food found in a snake was a 130lb (59kg) impala inside a 16ft (4.87m) long rock python, although there is a claim of a sloth bear weighing 200lb (91kg) inside an Indian python.

Despite the discovery of snakes having eaten such formidable meals, modern biologists have felt it unlikely that a large snake could take a fully grown human. A constricting snake must first catch its prey and then throw coils around the body to squeeze the life out of it. First, the chances of a healthy human being caught is improbable, although someone injured, resting or asleep would be vulnerable. Secondly, the lifeless body would have to be eaten head first and, despite a snake's ability to dislocate its jaws and swallow prey wider than itself, the shoulders would probably jam in the snake's mouth. If it did succeed, however, it would then use the coils to help work the prey down, and peristaltic movement of the gut walls would help slide it inside. The entire process would take a few hours.

Yet although adult humans are not prime targets for constricting snakes, children and babies certainly are, and to prove it there have been several incidents. On the Indonesian island of Selebaboe, one of the Talaud group of islands, a 14-year-old boy was consumed by a 17ft (5.2m) long python. When the snake was caught, killed and cut open two days after the attack, the boy's body was still intact and recognisable. It had not been crushed and the bones had not been broken.

In eastern Pakistan, an eight-year-old boy was taken by a 20ft (6.1m) python while he was gathering rice in a paddy field. The attack took place near Cox's Bazaar in April 1960. In 1972, a 14-year-old boy was taken from a paddy field in Burma. The local villagers caught and killed the snake, and after removing the victim's body, ceremonially cooked and ate the serpent in revenge.

In November 1979, a python in the Transvaal ate a 13-year-old South African shepherd boy. In October 1982, a 14-year-old herdboy from Richmond, South Africa, however, did not let a 16ft (4.9m) long rock python get the better of him. He bit it to death after it attacked him!

But could the biologists have got it wrong? If a modest-sized snake can swallow an impala or a crocodile, could a monster specimen swallow a person whole? In October 1998, a 24ft (7.3m) constricting snake was killed near an Indonesian village. When it was cut open, a large bulge part-way down its body turned out to be a man. The snake had swallowed him whole. There have been several other horrific reports.

During Abel journey in Tasman's Indonesia, for example, a certain Captain Ross was on his ship off the coast of Celebes when two Malays in a canoe came alongside. In their boat were the mangled remains of a man. His bones were broken, and his eyes were projecting from his head as if the head had been compressed. Ross asked the two men what had happened. They told him how they had landed their fish along the shore and had left the canoe in the charge of the poor man whose body he now saw. They told him to beware of large snakes at the edge of the wood, and left him. A few minutes later they heard his cries. When they reached the canoe, the man was being squeezed in the coils of a large boa constrictor. By the time they killed the snake, its victim was also dead. They showed the snake's head to Ross, and he was surprised how small it was. The snake's body could be no more than 8in (20cm) across at its widest part, yet the two Malays were certain the snake would have swallowed the man if it had not been disturbed.

There is also the case of a woman taken by a constricting snake claimed to be over 30ft (9.2m) in length. The report comes from Indonesia, where the island of Celebes, like Komodo, was dominated not by tigers and other mammalian predators, but by giant reptiles – enormous reticulated pythons. They feasted on the now extinct pygmy elephants and their young, but in more recent times have had to rely on domestic and feral animals, and on the occasional human.

The most frightening story of recent years comes from south-west Sumatra where a bulldozer of a forest logging project disturbed two giant snakes. According to newspaper reports, the driver of the machine had a life and death struggle with the two writhing monsters before he was able to crush one. It was claimed to be 82ft 6in (25m) long and inside its body were the bodies of four people. Two must have been taken not long before the snake was killed for their shorts were still intact.

You do not have to be in the jungle to end up as snake fodder. In August 1984, a pet python escaped from its cage in Ottumwa, Iowa, and strangled its owner's 11-month-old baby lying in his crib. A 15ft (4.6m) long pet python killed its owner, an American man who had tried to handle the creature unassisted. A 21-year-old American woman found herself with her arm being pulled down the throat of her 12ft (3.7m) long pet python when she went to clean its cage. Fortunately someone was able to prise the snake's mouth open and she escaped unscathed. In August 1996, a 9ft (2.7m) long Burmese python (*Python molurus bivittatus*) – a sub-species of rock python popular with amateur herpetologists – wrapped itself around a pregnant women in a San Diego hotel room and began to squeeze the life out of her. The husband wrestled with it for 20 minutes or more, but could hardly budge it. Staff called for the fire brigade and a fireman eventually hacked off the snake's head. The snake was the husband's pet: he described it as a 'slippery puppy', but vowed never to keep snakes again. And in October 1996, a Burmese python was in the headlines again when a 12ft (3.7m) long specimen squeezed the life out of a 19-year-old boy in New York. The boy had been preparing a chicken to be fed to the snake when it went for the larger prey. Police recorded an accidental death.

The most bizarre snake story, however, comes from China. A Chinese hunter and his brother were returning home to the village of Li Bian in Huang County, Shanxi province, when they chanced upon a snake. The man pinned the snake down with the butt of his gun, but the snake coiled itself around the gun; its tail struck the trigger and shot the man in the buttocks. He died on the way to hospital.

# CHAPTER 11

---

# SHARK ATTACKS

There is no doubt that sharks touch that inner fear which most of us have at one time or another of unknown creatures from the deep. How often have you peered down into the inky darkness of the ocean and wondered what was lurking there ready to strike? Psychologists suggest that this fear of dangerous animals stems from humankind's shadowy and distant past, when our ancestors stepped down from the trees and left the shelter of the forest and were confronted with the most terrifying of threats – that of being eaten alive. Sharks, more than any other man-eaters, bring out this ultimate of fears. Sharks terrify yet intrigue us in a way no other animals do.

Of the 390 known species of sharks, only a handful attack and eat people, and even these are not thought to target humans deliberately as prey. If they did, then, given the number of people in the water at any one time – bathing, surfing, canoeing, wind-surfing, body-boarding, skin-diving – there would be far more shark attacks reported each year. The reality is that shark attacks are relatively rare. Analysis of data in the International Shark Attack File (the repository of all known shark attacks worldwide) reveals that there are no more than 20 to 50 recorded attacks each year, and of those only 10 per cent are fatal – at worst about 10 attacks per year. In fact, during the period from 1990 to 1996, the average number of recorded shark attacks on people was 49 per year, six of which were fatal. When unrecorded events are included, from parts of the world where statistics are not kept or news of attacks is suppressed for fear of bad publicity, the number of attacks worldwide is estimated to be no more than 70–100 annually.

Recently, there has been a marked increase. The number of shark attacks on people worldwide, according to a University of Florida report, increased substantially during 1997. While there were 36 unprovoked cases reported in 1996, the figure had increased to 56 in 1997. The highest number in recent years was 72 in 1995. Off South African coasts, shark attacks quadrupled during 1998, compared to the previous year. Putting this into perspective, however, there are more fatalities from automobile accidents around the world in one month than fatal shark attacks on people during recorded history. A swimmer is far more likely to be killed on the way to the beach by a car than in the water by a shark. In fact, it is thought by some authorities that the behaviour of people, say, on a beach, is more closely related to the pattern of shark attacks than the behaviour of sharks.

In a review of the International Shark Attack File in 1974, David Baldridge revealed that although attacks off a beach can occur in any depth of water, most take place in water which is just waist deep or less. This does not mean that this is the most dangerous place to be: it is simply the place at the water's edge where most people are playing, swimming or just standing. The real risk of attack, however, increases as a swimmer moves further from the shore: the deeper the water the more the victim is in the shark's domain. He found that more men are attacked than women, in a ratio of about 8:1. The greater number of attacks on men appears to be linked to the way in which the human male behaves at the beach. Men are not only more likely to be involved in aggressive, active behaviour, such as splashing and fooling around, but they also swim the furthest from the beach and are most likely to be on the outer fringe of a group of swimmers. In short, men are more accessible to a passing shark. The observations gain even more credibility when the records examined are limited to incidents outside of a line 130 yards (120m) from the shore. In this zone the male to female attack rate increases dramatically to 30:1.

Yet although the risk of attack increases the further a swimmer is from the beach, the actual number of attacks is highest closer to shore. Over half of all attacks occur within 66 yards (60m) of the tide line, the zone coincidentally where most bathers are to be found. In April 1987, for example, the arm of a 16-year-old girl was bitten off by a shark in chest-deep water at Mustang Island, Texas. On 14 September 1988, a man was killed by a shark in just 8ft (2.4m) of water off the jetties of St Andrew Pass, near the US Tyndall Air Force base, and about half-an-hour later a couple of tourists were injured by a shark – possibly the same individual – in only 4ft (1.2m) of water. The water had been particularly murky.

During the summer of 1995, 25 attacks on people occurred along Florida coasts, including surfers who had limbs bitten and a diver killed by a bull shark (*Carcharinus leucas*). A succession of hurricanes brought large waves and murky waters to the coast and local researchers feel that surfers and swimmers were bitten in cases of mistaken identity. In the cloudy water, the sharks were pursuing schools of fish, but when confronted with a person moving in the water, they did what comes naturally. The 32-year-old son of a British royal was attacked in knee-deep water off a Florida beach in November 1997. Thirty stitches were required to close the wounds.

In December 1998, a nine-year-old boy was some 40 yards (40m) offshore in about 10ft (3m) of water off Vero Beach, Florida, when a tiger shark (*Galeocerdo cuvier*) dragged him under before anybody could get to him. His body, with head and arms missing, was found the following day. It was only the second recorded fatality in ten years along the Florida Panhandle.

Victims can even be in water just a few inches deep, some of the more bizarre attacks taking place right at the water's edge. In August 1966, for example, an eight-year-old boy and his mother were paddling in about a foot (30 cm) of water at Rivera Beach, Florida. The sea was rough, the result of a violent offshore storm and the boy was enjoying the waves. His mother was about 2ft (60 cm) from him, when, out of the corner of her eye, she saw a grey shape heading straight for her son. She grabbed his arm and lifted him bodily out of the water. At that moment a shark shot underneath the suspended boy and beached itself on the sand. There it thrashed about until it was able to wriggle into water deep enough for it to manoeuvre itself back out to sea. The boy and his mother continued along the shore, keeping well clear of the water, while several sharks swimming in the surf followed their progress.

A similar event took place in February 1972 at Taperoo Beach, near Adelaide in South Australia. On this occasion a 6ft (1.8m) long unidentified shark tried to attack a woman in shallow water. She ran and the shark charged. She escaped and just kept running up the beach, but the shark could not stop and ended up high and dry on the sand. The coastguards were summoned and the shark was dispatched with a hammer. In January 1986, a New Zealand doctor stunned an unidentified shark which was approaching his children on the beach at Invercargill, at the southern tip of South Island. He hit it with a cricket bat! In October 1986, a mother prised open the jaws of a 10ft (3m) long shark in order to release her seven-year-old daughter. This took place at Sanibel Island, off Florida's Gulf coast to the west of Fort Myers.

More recently, on 22 July 1996, a man wading in water just 2–4ft (0.6–1.2m) deep at Cold Storage Beach, at Cape Cod, was attacked by a blue shark (*Prionace glaucus*) estimated to be 6–8ft (1.8–2.4m) long. It is thought the shark was accompanying tuna that had been driving bluefish towards the shore. The shark mistook the water disturbance caused by the man's legs for the flapping of its prey. The man hit the shark repeatedly in the face until it let go, and before the shark could attack again he was able to get clear of the water. It was the first shark attack in the area for 11 years.

David Baldridge and his colleagues went further with these studies and, analysing data outside the International Shark

Attack File, observed the way in which people actually behaved on beaches. Counting bathers at Myrtle Beach, South Carolina, they noticed that the distribution of bathers strongly tied in with the distribution of known shark attacks. About 17 per cent of bathers waded in knee-deep water which, surprisingly, matched the statistics for shark attack victims in that depth of water: about 16 per cent. About 73 per cent of bathers swim in water up to neck deep, and here again the match is remarkable: 78 per cent of attack victims were swimming in water in this zone. The depths at which shark attacks occur seem to be related to nothing more than human population distribution on the beach and to no significant behavioural or physiological reason on the part of the shark.

Temperature is another factor often considered when shark attack statistics are analysed. It has been suggested by several authorities, for example, that the sea temperature must be over 70 degrees F (20 degrees C) before a shark will strike a person. Looking at the beach study, however, researchers were able to come up with a simple explanation. Sure enough, they found that people are reluctant to go swimming in water below 70 degrees F (20 degrees C). Below this temperature there would be fewer people in the water for a shark to attack. As the temperature rises so does the number of swimmers entering the sea. Once again it is the behaviour of people rather than the behaviour of sharks which seems to explain the pattern and frequency of shark attacks on people.

At one time, one of the most dangerous beaches for shark attacks was Amanzimtoti, a holiday resort about 17 miles (27km) to the south of Durban on South Africa's Natal coast. It possesses the unenviable distinction of having one of the highest records of shark attacks of any beach in the world. Two young men were attacked and killed by sharks there in February and December 1940, and a 17-year-old was a victim in December 1943. Then, there was a lull until 3.30 in the afternoon of 30 April 1960. A 16-year-old boy was in 10ft (3m) of water, just 33ft (10m) from the shore when he felt what he thought was seaweed wrapping around his leg. In fact, it was a shark bite, and the attacker did not stop at that. The shark returned, grabbed the boy and dragged him below the surface.

The boy struggled and, while trying to feel for the shark's eyes, caught his fingers between the teeth. The skin was stripped off the index finger like peeling off a glove. The shark let go of the boy momentarily and he was able to head for the surface, but the shark returned. It bit into his abdominal wall. The boy felt no pain, but had the overwhelming desire to live and struck out for the shore. The life-and-death struggle had gone on unnoticed from the beach, but, as the boy emerged from the water with his intestines hanging out of his abdomen, people on the beach were galvanised into action. Lifesavers carried him to their hut, a doctor arrived quickly and administered first aid, and the boy was taken to the local hospital. Miraculously, he survived.

The next attack was on 22 January 1961. A 15-year-old boy was swimming in the surf in no more than 6.6ft (2m) of water when a shark shredded the flesh on his leg from the thigh to the calf. The injury could not be repaired and so the leg was amputated, but the boy survived. First two and then three shark nets were placed offshore to reduce the risk of shark attacks, but just after midday on 17 April 1963, the sixth shark attack occurred. Two 15-year-old boys were swimming in water about 100ft (30m) deep about 131ft (40m) from the shore, when a shark bumped one of them, removing the skin and surface muscle from his left buttock. The two swam for shore, and the victim was taken to hospital. Another net was added to the shark defence installation, and ten years were to pass before there was another spate of attacks.

On 7 January 1974, a professional lifeguard was about 328ft (100m) from the shore when a shark grabbed his knee. He punched his attacker and it let go. On 13 February 1974, a 14-year-old boy was bodysurfing in water 5ft (1.5m) deep just 16ft (5m) from the shore when a shark grabbed his right leg. It shook the boy several times and then left. The boy was able to struggle on to the sand, his calf muscle shredded and blood spurting everywhere. Lifesavers cared for him until an ambulance took him to hospital. On 21 March 1974, a 21-year-old boy had a narrow escape when a shark attacked but only bit into his surf board. On 4 April 1974, a 17-year-old surfer needed 19 stitches in his foot after an attack, and on 23

February 1975 a surfer lost his left foot to a shark. The last four attacks took place on days when bathing was banned officially because the authorities believed that conditions were conducive to shark attacks. The attacks also coincided with the build-up of fish populations in the area, triggered by events further down the food chain, such as the annual seasonal migration along the South African coast of anchovies and sardines. It is this abundance of food along South Africa's coasts that attracts such a diversity of marine creatures, including many different species of sharks.

For the holiday resort of Karridene on South Africa's Natal coast, shark attacks not only terrified the entire community but also brought it to the edge of bankruptcy. The year was 1957, the month was December. It was to become known as 'Black December', and the horror started at 5 o'clock in the afternoon of the 18th. A 16-year-old amateur lifesaver was bodysurfing in chest deep water, no more that 164ft (50m) offshore. He felt what he thought was seaweed brush his leg, but then saw a shadow and, a few seconds later, a fin as it broke through the surface. The shark turned and grabbed his leg, pushing the boy through the water. He punched and struggled, but the shark had bitten through his leg at the knee. Nevertheless, he survived.

Two days later, at 4 o'clock in the afternoon of the 20th, a 15-year-old boy was standing in the surf on a sandbank about 100ft (30m) from the shore at nearby Uvongo, when a shark pushed him backwards. All witnesses saw was the tail of a large shark 'lashing the water'. The boy was dead before his badly mutilated body could be recovered. Three days after this incident, a 23-year-old man was bathing with many others at Margate, when people on the shore spotted the dark shape of a shark heading towards them. Before a warning could be shouted, the shark took the young man and friends saw him 'wide-eyed, ploughing sideways through a breaking wave'. The two twisted and somersaulted in the waist-deep water, as the shark pressed home its attack. When companions reached his body his left forearm had been bitten clean away, the right arm had been stripped of flesh, the lower abdomen, buttocks and right thigh had been eaten. He did not survive.

Holidaymakers began to leave Natal's south coast, but a few remained. At 1 o'clock in the afternoon on 30 December, a 14-year-old girl was with a small group of bathers in waist-deep water slightly removed from the many hundreds who packed Margate beach when, without warning, a shark grabbed her left buttock and dragged her under. Attacking a second time, it bit off her left arm at the shoulder and sliced into her left breast. As the sea reddened with blood and the other bathers watched in horror, a brave young man raced to her rescue. He grabbed the shark by the tail and, as he tried to wrestle it off, it almost casually flicked him away. The man did not give up, and eventually managed to prise the girl away from the shark. She was carried to the beach and whisked away to a hospital unprepared for such a badly mutilated person. During the six-hour operation, staff and volunteers gave blood, and the girl was saved.

The beaches were closed instantly, and a shark-catching frenzy ensued. Even the South African Navy was drafted in, the minesweeper *Pretoria* dropping depth charges in an attempt to kill sharks along the coast. It was all to no avail, of course, for on 9 January 1958 a shark rammed a 42-year-old man in thigh-deep water, sending him flying into the air. Turning rapidly, it attacked again and again, removing portions of both thighs and large chunks of buttock. Lifesavers ran into the water, and rescued the man but he died before he reached hospital.

Panic set in and people left the coast in droves. The local authorities erected a few unsightly barriers in the sea in the hope that people would come back at Easter. They did . . . and so did the sharks. At 2 o'clock on the afternoon of 3 April a 29-year-old male holidaymaker, with mask and flippers, was swimming in chest-deep water about 66ft (20m) beyond a group of bathers in shallower water at Port Edward. Suddenly, and without warning, he was repeatedly attacked by a shark. Another man, showing great courage, ran into the surf and dragged the victim back to the shore. The onlookers were horrified. The dead man had both arms and a leg severed, and a large part of his abdomen was missing.

Two days later, very low spring tides left the shark barriers at

Uvongo clear of the water, and so the local authority repaired any damage. A small group of people gathered at the mouth of the Uvongo River to watch the workmen. Among them was a 28-year-old mother of four children whose husband had been killed in a motorcycle accident not long before. The group stood in knee-deep water, and as they stood there a shark swam rapidly out of a channel and knocked the woman into the water. It turned, clamped its jaws around her middle, and almost bit her in two, killing her instantly. By midday, the news had spread and there was a mass exodus not only from the water, but also the entire holiday resort area. In an extraordinary outburst of mass hysteria, people just upped and left *en masse*. Cars blocked roads for many miles around. The resorts were deserted, and the people did not come back. Many hoteliers and small businesses went bankrupt. The tourist industry was wiped out.

Although the number of attacks, even in these high risk areas, is relatively small, say, compared to the victims of automobile accidents, this is no consolation for shark attack victims and their families. The plain truth is that, although attacks are few and far between, sharks do attack people and the circumstances are often horrific. Some species of sharks are deadly. They are in an environment in which they are totally and utterly at home, a habitat in which we are simply clumsy and vulnerable newcomers and therefore fair game for a highly efficient predator.

In another study of the International Shark Attack File, Leonard P. Shultz focused on 1,406 attack reports dating back to the mid-1800s. He found that half the attacks occurred on swimmers at the surface or people wading in shallow water from knee-deep to chin-deep; attacks occurred in all weathers – cloudy, sunny or stormy; attacks were in clear and murky water, in daylight or in the dark, in the open sea, in shallow seas, in river mouths or in narrow rivers leading to the sea; and attacks were in waters of all temperatures. In short, sharks may attack few people, but when they do so they attack at any place, at any time, and under any conditions.

'In our experience,' pioneer underwater explorer Hans Hass once said, 'their teeth are among the most terrifying murder instruments in the Animal Kingdom. A twelve-feet shark can cleanly bite off an arm or a leg; a twenty-two feet shark can bite a human body in two.' And, Jacques Yves Cousteau once wrote, 'I and my diving companions fear them, laugh at them, admire them, but are forced to resign ourselves to sharing the waters with them.'

They are sentiments shared by mariners down the ages. One of the earliest recorded shark attacks took place in 1580, and was observed by horrified sailors attempting to rescue a colleague who had fallen overboard on a voyage between Portugal and India. The man was thrown a rope, but as the crew pulled him to 'within half the carrying distance of a musket shot' of his ship, a large shark appeared and tore him apart before they could get him aboard. A full account of the event, in a letter from an eyewitness, is lodged with the International Shark Attack File, as is another which took place in 1595. It occurred at the 'River of Cochin' on the south-east coast of India, to the north of Allepy. A sailor was lowered over the side to fix the ship's rudder in place, when a shark came up and bit off his leg cleanly at the thigh. The captain struck the shark with an oar, but it kept coming back. The victim also lost a hand and arm above the elbow and a piece of his buttock.

A historical account from 1749 tells how a 14-year-old orphan called Brook Watson had his right leg removed below the knee by a shark in Havana Harbour, Cuba. Watson survived the attack, and eventually became an MP and Lord Mayor of London. His amputated leg formed part of his coat-of-arms when he became a baronet.

Another attack was described in Thomas Percy's *Bryan and Pereen: a West Indian Ballad*:

> Then through the white surf did she haste
> To clasp her lovely swain
> When ah! a shark bit through his waist:
> His heart's blood dyed the main!
>
> He shrieked! he half sprang from the wave,
> Streaming with purple gore,
> And soon it found a living grave,
> And ah! was seen no more.

Ancient and modern, all those who have witnessed, observed or studied shark attacks, have recognised from the statistics that any shark 6ft (2m) or longer is potentially dangerous to a person in the water, and four species in particular have been implicated in serious attacks on people. They head the shark attack league tables. They are, not surprisingly, the great white shark (*Carcharadon carcharias*), the tiger shark (*Galeocerdo cuvier*), the bull shark (*Carcharias leucas*) and the oceanic white-tip shark (*Carcharhinus longimanus*).

Whether great whites are always the perpetrators of attacks attributed to them is another thing. The identity of an attacker is not easy to establish accurately in the heat of the moment. The only sure cases are those in which fragments of tooth have been left behind. Tiger sharks and bull sharks are thought to be responsible for many attacks attributed to great whites. They are numbers two and three on the shark attack league table.

The tiger shark is probably one of the most dangerous sharks with which people will come into contact regularly worldwide. It is certainly considered to be the most dangerous in tropical waters. Attacks on people have been recorded in all tropical and sub-tropical seas. A 20ft (6m) long tiger shark, for example, was thought to have been the villain in an attack on a woman in Hong Kong in June 1991, the first there since 1979. And, whenever tiger sharks are implicated in attacks, the events are sure to be disturbing.

On Sunday, 24 July 1983, the 40ft (12m) long fishing trawler *New Venture* foundered off the Great Barrier Reef in Queensland, Australia. A boom broke, the boat heeled over, was hit by a freak wave, and turned turtle, throwing the skipper and his crew – a fellow fisherman and a female cook – into the water. At first they clambered on to the upside-down hull but gradually the trawler filled with water and sank. The crew grabbed a life-ring, three large pieces of foam, and a surfboard to keep themselves afloat, confident that they would be spotted and rescued.

On the evening of the following day a shark appeared – a 15ft (4.5m) long tiger shark. It circled them, nudged the life-ring and pieces of foam but did not attack. At this point the crew was not particularly worried: the three had seen sharks before and hoped it would just lose interest and go away. But it did not go away. Instead, it came up under the surfboard and tried to bite the skipper on the knee. He kicked at the shark and it stopped its attack. Five minutes later large waves knocked the three shipwreck victims into the water and the shark immediately grabbed one of the crew by the leg and bit most of it off. The crewman then did a brave thing: knowing he was done for, he swam away from his companions, hoping that he would lure the shark away from them. The shark attacked and killed him, dragging the body below.

The skipper and his surviving crew member paddled for all they were worth towards nearby Loader's Reef, but the shark returned. It grabbed the cook, shook her four or five times like a rag doll, and dragged her below. The last survivor – the skipper – paddled again for the reef. The shark returned yet again. Every time it came close by, the fisherman stopped paddling and let himself drift until it had gone. After what seemed hours he eventually reached the reef, and a spotter plane flew over. He was about 100 yards (90m) from the surf at the edge of the reef when he saw the shark zigzagging towards him again. Fortunately he was swept by a wave over the reef and able to scramble ashore. At first he laughed hysterically and then broke down and cried. It was undoubtedly one of the most horrific shipwreck and shark attack stories in Australian history. It became front-page news in Australia, and was widely covered in newspapers and on radio and television all over the world.

Tiger sharks are not only hunters, but also scavengers – the ocean's dustbins, according to some shark scientists. They frequent inshore waters close to slaughter houses where they feast on parts of sheep, pigs, cows and horses that have been dumped into the sea. In 1935, this scavenging habit placed a 14ft (4.2m) tiger shark centre stage in a murder inquiry in Australia.

The shark was caught in a fishing net and, as it was still alive, it was given to the Coogee Aquarium in Sydney. It only survived for a week, but during that time it regurgitated its stomach contents, which included a muttonbird and a human arm attached to a piece of rope. The arm was decorated with a recognisable tattoo, and was identified as that of a missing person, in fact, one James Smith, a member of the Sydney underworld. He had been part of a gang of fraudsters, but they had fallen out over a bungled job. Smith was eliminated, according to one of the gang who later committed suicide, and his body stuffed into a metal box, all, that is, except one of his arms which was cut off, tied to a weight and thrown separately into the sea. The story would never have come to light if the captured tiger shark had not swallowed the arm.

The Sea of Cortez (Gulf of California) is host to tiger sharks – big ones. Sharks estimated to be 10–15ft (3–4.6m) long patrol the sea-lion colonies at the north end of islands in the gulf. Mexican divers are reluctant to swim near the north end of any of the gulf islands, such as Santa Catalan, Cerralvo, Carmen and San Marcos, for fear of a 'mistaken identity' attack. The sea-lions clear the water immediately the tiger sharks appear; anybody caught up in the mayhem is fair game for the sharks.

Tiger sharks have also being operating in the vicinity of Cozumel off the north-west tip of the Yucatan, a popular diving destination for tourists, if reports in 1997 are to be believed. The first sign that something was not right occurred on 15 August when two divers were thought to have been attacked by a tiger shark at the Santa Rosa Wall. One of the bodies was recovered. It had an arm and a leg missing. The other body was never found.

Similarly, Hawaii became a focus of world attention in recent years, and some of the attacks are thought to be cases of mistaken identity. In 1958, a boy playing on an airbed at Lanikai, on the east coast of Oahu, was killed by a shark thought to be a tiger shark. The event triggered an overwhelming desire for revenge among the people of Oahu and they slaughtered the local shark population. Of the 500 or so sharks killed, 71 were tiger sharks, indicating that this species constitutes a relatively high proportion of Hawaii's shark population.

In 1985, a body-boarder (boogie-boarder) was attacked by a tiger shark off Princeville, Kauai. There were several sea turtles in the area at the time of the attack. In August 1996, a boy was body-boarding off Ukali Road, Paukukalo, Maui, and was attacked on the calf by an 8ft (2.4m) long shark. The boy punched the shark several times before it let go. And in October 1997, a tiger shark was implicated in an attack on another body-boarder. The boy was about 150 yards (137m) offshore in 5–6ft (1.5–1.8m) surf in Waiokapua Bay, Oahu, when a shark – most probably a tiger shark, thought to be about 10ft (3m) long – bit off part of his right leg at mid-calf. His hands and left foot were also lacerated during the attack. A tourniquet, made from a body-board leash, was tied around the thigh to stem the flow of blood and the boy rushed to hospital. He survived. The time was about 7 o'clock in the morning, and the water was reported to be turbid. The head of the shark was estimated to be about 2ft (0.6m) wide, and the shark itself about 13–14ft (4–4.3m) long. Interestingly, several of the attacks, including those on surfers in March and October 1992 were early in the morning. Some surfers are always out before the sun rises. Dawn and dusk are known activity times for sharks.

Since December 1993, shark attacks on people around the Hawaiian islands have been fewer. This period of relative calm followed a spate of attacks that occurred for an uncomfortable 33 months from April 1991, when 13 people became victims, including four deaths. It is thought that these and other shark attacks in the area could well have been cases of mistaken identity. While a surfer on a malibu surfboard resembles a seal seen from below, and therefore attractive to great white sharks, a person on the smaller body-board bears a striking resemblance to a sea turtle and is therefore likely prey for a tiger shark. The attack at Princeville, for example, involved a body-board with a yellow-coloured underside, resembling the plastrom (undershell) of a sea turtle.

Merely swimming in a known sea turtle area could be enough to invite an attack. On 17 January 1996 a male swimmer with face mask and snorkel was bitten on the foot and lower leg by a shark estimated to be 6–8ft (1.8–2.4m) long. He was swimming at Napili Point, West Maui, an area known to be frequented by turtles. The shark was described as a 'wall of grey', with a tail in which the upper lobe was longer than the lower lobe and curved. This suggested to the local experts that the attacker was a tiger shark, most likely on the lookout for sea turtles. West Maui was also the focus of a shark alert in March 1999, when, at about 11 a.m., a 29-year-old woman was bitten on the thigh by a tiger shark just 300 yards (274m) off Whalers Village condominium in Kaanapali. It is thought an injured humpback whale could have attracted tiger sharks to the area, but the shark took a large bite out of the right upper thigh of the woman rather than the whale. The size of the wound indicated a shark about 12–13ft (3.7–4m) long. The girl and her boyfriend were out diving with the whales, when the crew of the charter boat – the 64ft (20m)) catamaran *Gemini* – heard their cries for help. The girl's companion had kicked the shark in the head to make it let go, and the *Gemini* came to the rescue. When she was brought aboard, she was conscious and breathing, and her leg was wrapped in towels. An ambulance crew was waiting at the shore. They discovered that muscle and skin were missing from above her knee in a gash about 15in (38cm) long.

A few days previously, an 18-year-old man was attacked while body-boarding at Kealia, Kauai. He had a large gash on his right calf but survived. And the previous October a visitor had been bumped on a diving tour south of Lahasina, in waters off Olowalu. During the same month, two other attacks occurred in the same area. The most horrific was that of a newly married couple who had rented a kayak but were swept out to sea to the south of Lahaina by strong trade winds. As night fell, it was warmer in the water than on the wind-swept craft, but while they were waiting for rescue a tiger shark arrived. It ripped the arm off the wife and she died from loss of blood. She slipped into the water, while the husband clung to the kayak and was washed ashore 12 miles (19km) away on Kahoolawe, an island used by the US military for target practice. Eventually he was rescued and returned to Maui. The third attack that month was off Kauai. An 18-year-old boy was attacked while body-boarding, but suffered only minor injuries. And, in July 1999, a shark with a maw about 16in (41cm) across took a bite out of the buttock and right thigh of a 43-year-old man. The attack took place at about 10 o'clock in the morning at Honolii Bridge, near Hilo on Big Island. Local medical centre staff thought the man was lucky to be alive.

If sharks were deliberately seeking out people, however, then there would be far more attacks reported; after all, Hawaii's beaches are filled daily with bathers and surfers: but the sharks do not usually bite people.

Exceptions to the rule, however, are bull sharks. These streamlined but heavy-set members of the requiem shark family are unusual among sharks in that they enter fresh-water. As a consequence, many attacks on humans occur in the turbid waters of large river estuaries. The most horrific case took place many years ago, but the events are still debated today. Centre stage is thought to have been the bull shark. This gruesome series of attacks took place on the US east coast in July 1916. Ironically, while hundreds of thousands of soldiers were being killed in Europe during the First World War, five shark attacks within ten days made front-page news worldwide. All took place along the New Jersey coast, and the most horrific occurred at Wyckoff Dock, a popular bathing spot in Matawan Creek.

The fateful day – 13 July 1916 – is etched on the minds of all those who lived in New Jersey at the time. Already, there had been two attacks at the coast. On 1 July a 25-year-old man was swimming just 16 yards (15m) from the shore when people on the beach spotted a shark's dorsal fin cutting through the water and heading straight for the swimmer. They shouted for him to get out of the water and he struggled to make it to the beach, but the shark took him and pulled him below. Five days later a young man was swimming outside the life-lines at Spring Lakes when he too was attacked and killed. A week later, it was Matawan's turn to face a tragedy. In the morning of the 13th, a retired sea captain was crossing Matawan Bridge when he peered down into the water and spotted the dark shape of a shark swimming along with the incoming tide. He ran immediately to the town's barber who was also the chief of police, and then shouted a warning to people heading for the dock to bathe. 'There's a shark in the creek,' he called, but nobody believed him. Despite the previous attacks and regardless of the old man's warnings earlier in the morning, several boys jumped into the murky water to cool down in the heat of the day. Suddenly, one of them felt something abrasive brush against his midriff and he left the water with his stomach streaked with blood. 'Don't dive in any more – there's a shark or something in there!' bystanders called out. But, the warning was ignored and as one of the group shouted, 'Watch me float, fellas!', another boy felt and then saw a huge fish slam against his legs. Before he could shout a warning the other boy, who was the furthest from the pier, was grabbed, reappeared, screamed and then disappeared in an instant.

At first eyewitnesses thought the boy had had a fit, but then the reality dawned. A hungry shark had entered the narrow creek, just 11 yards (10m) wide, and it was dangerous. Would-be rescuers packed into the creek, some in boats and others diving into the murky waters not realising what had happened and what fate might be waiting for them below the surface. Indeed, one brave rescuer dived several times into the deeper part of the creek looking for the boy who had been taken. He made contact with the body and was dragging it to the shore when he too was attacked. A stain of blood appeared around him, according to eyewitness reports, and he emerged from the water holding his severed right leg above his head. He died later in hospital. A little later and 867 yards (793m) downstream another victim, a 14-year-old boy was hit. News had just arrived that there had been two shark attacks at Matawan, and he was trying to climb the ladder up the side of the dock when he was struck. The shark grabbed his right leg. He later told how he could feel it going down the shark's throat. He was rushed to hospital and survived.

The local community was up in arms, quite literally. Men appeared with hand-guns, rifles, shot-guns and even sticks of dynamite. Several sharks were caught near the creek, but attention focused on a 8ft 6 ins (2.6m) long great white shark. In its stomach were the tragic remains of human victims – 15lb (7kg) of human flesh, a child's shin bone, and a rib bone. Whether a great white was responsible for the attacks or it had simply scavenged the body parts is still debated today. Experts believe that a bull shark – a species more usually associated with brackish, estuary waters, and a known man-eater – was the shark responsible for the fatal attacks and that the great white – a small one, as it happened – was an innocent bystander that took advantage of an easy meal. Nevertheless, it was caught in Raritan Bay, New Jersey, and stuffed by a taxidermist who put it on display. Over 3,000 people paid ten cents each to stare at the alleged man-eater. The bull shark – probably the real killer – escaped.

The bull shark is number three in the league table of sharks known to kill people, and is extremely aggressive, both in the sea and in rivers and lakes. Indeed, it is regarded by some shark experts to be the most dangerous of all sharks. It is thought to have been responsible for many attacks originally attributed to great whites, particularly in warmer waters. It is as catholic in its tastes as the tiger shark, but it has powerful jaws, wider in relationship to its body length. It has a blunt snout, small eyes, triangular-shaped teeth in the upper jaw with coarse serrations, and dagger-like teeth in the lower jaw. The upper surface of the body is grey, and the edges of the fins are dark in younger specimens. The shark is a first-class hunter, with refined senses including the ability to hear sounds in the 100-1500 Hertz range, the middle part of the spectrum of human hearing. It cruises relatively slowly but the muscles of its heavily built body can put on an extra-ordinary burst of speed when homing-in on a target.

The shark's habit of patrolling inshore waters also brings it into more frequent contact with people. Attacks in the Zambezi and Limpopo rivers, for example, have been attributed to bull sharks, and in the sacred Ganges the bull shark is thought to be responsible for attacks previously attributed to the very rare Ganges shark (*Glyphis gangeticus*). It has been known to frequent the bathing ghats of Calcutta, feeding on partially cremated bodies and attacking religious bathers. In one year, as many as 20 pilgrims were attacked, half of whom were killed. In the estuary of the Devi, five people were killed and another 30 badly mauled during a period of two months in 1959. In the estuary of the Limpopo in Mozambique, three attacks on people occurred within a space of six months during 1961, all attributed to a single rogue bull shark. One attack was 150 miles (241km) from the sea.

During the Second World War, a most unusual attack took place at Ahwaz in Iran, about 90 miles (145km) from the sea. A British soldier, intent on washing the mud from his ambulance, stepped down into no more than a foot of water in the Karun River (a tributary of the Tigris) and, as he was beginning to set to work, was seized on the leg by a shark. Caught off balance, he fell over and began to fight for his life in water no deeper than in a bathtub. Defending himself with arms and fists, the soldier was badly lacerated. His right arm was torn open, his hands slashed, and his leg badly gouged. He was one of 27 people attacked by sharks between 1941 and 1949, the period when Allied military authorities kept records. All had been attacked in shallow water. Bull sharks were the likely culprits.

Local folklore has it that the sharks wait below date palms at Khorramshahr, downstream from Ahwaz. In the Tigris, they are supposed to head for Baghdad and its melons, 350 miles (563km) from the sea. Folklore aside, many rivers entering the Persian Gulf have been the sites of shark attacks. Danish biologist H. Blegvad, writing in a paper on the fish of the Persian Gulf, stated that many people, especially children who play in the rivers, were killed by sharks every year. He thought that with less food available in rivers than in the sea, sharks that enter freshwater were more liable to attack people. It would be interesting no know whether these types of shallow-water attacks still occur today in the rivers at the head of the Gulf.

Some Australian east coast rivers are known to be visited by sharks. The Brisbane River in Queensland and George's River, New South Wales, have been the locations of shark attacks. The stories are not recent but nevertheless relevant. In the Bulimba Reach of the Brisbane River in 1921, a man was carrying his young son on his back and wading to his dinghy, just 30ft (9.1m) from the river bank. A shark seized his hip, and in the struggle the boy slipped off and into the water. The shark let go and the boy was taken, never to be seen again. At East Hills, about 20 miles (32km) from the mouth of George's

River a shark struck a 15-year-old boy swimming to fetch a tennis ball. The boy managed to get back to the river bank, only to turn around and find that the river was filled with sharks. Onlookers gazed in amazement. A year later, in 1935, another boy was swimming in a race with his companions, about 2 miles (3.2km) upstream from East Hills, when he was struck by a shark. He was dead before his rescuers got him to the river bank. On the same night, a further 3 miles (5km) up river, a young girl playing in about 4ft (1.2m) of water had both her hands bitten off by a shark. She survived by the prompt action of rescuers. In 1927, at Eden, New South Wales, a man on a horse were attacked by a shark in the Kiah River. The horse bucked and the man was thrown into the water. The commotion startled the shark and it swam away.

Whether these Australian attacks were by bull sharks must remain speculation, but the pattern of behaviour suggests this species is a likely candidate.

Central and South American lakes and seas are also haunts of the bull shark. In Lake Nicaragua, site of the first observations of freshwater sharks, it is a known killer, accounting for at least one human death and numerous dog deaths each year. In 1944, one shark attacked three people near Granada, the largest town on the lake. Two of them died. Of late, sightings of bull sharks in the lake have been few, the result of bounty hunting introduced by the authorities in Granada or by the silting-up of the San Juan River.

Like many sharks, the bull shark is quick to take advantage of changing circumstances and new feeding opportunities. Such was the case at beaches in Recife, north-west Brazil, when the ecology of the area was changed radically by human activity. On 28 October 1996 a surfer and body-boarder were attacked by a 8.2ft (2.5m) bull shark. At 10.30 in the morning the two surfers were in murky water, about 329ft (100m) from the beach at Barra de Jangada, close to the mouth of the Jaboatao River. One was seized on the knee and the other was badly bitten on the thigh when he went to help. Tourniquets made from the board-straps prevented them from losing too much blood and they both survived. A damaged tooth recovered from an injured leg was identified as from a bull shark.

These accidents represented the 22nd shark attack, with six fatalities, since September 1993 on the same 6.2-mile (10km) stretch of beach where no attack had been recorded previously. It occurred despite a ban on surfing in the area since January 1995. The sharks came in a variety of sizes, 3.3–10ft (1–3m), and five were positively identified – four bull sharks and a tiger shark. An increase in shipping to and from a nearby port was coincident with the sudden spate of attacks. A shark workshop attended by experts in November 1995 identified several other factors: there was a definite increase in the number of surfers and bathers in the area; the by-catch from shrimp trawlers was dumped nearby; the near-shore river channel is a favourite spot for bull sharks; and climatic change with more southerly winds and less rain at the time of the attacks drove ocean currents from south to north, the water moving from the new port area to the beach south of Recife. Again, human activities and people's behaviour have more significance in shark attacks than the behaviour of the sharks themselves.

The carnage, however, has continued. In November 1998, a surfer was attacked and later died after a shark had bitten off his hands at the Praia da Boa Viagem beach of Recife. It was the second fatal attack during 1998 and the 27th shark attack in the area in two years. And, in May 1999, Boa Viagem was to feature again when two sharks, one an 8ft (2.5m) long tiger shark, were thought to have attacked a 21-year-old male surfer. The sharks attacked several times, biting into the man's thigh and hands, but he was able to fight back and was rescued eventually by three lifeguards.

The fourth man-eater is the oceanic whitetip shark (*Carcharhinus longimanus*). It is easily recognised by its large spade-like anterior dorsal fin with the mottled white tip and the paddle-shaped pectoral fins. It is one of the larger species, with mature fish reaching up to 13ft (4m) long. Unlike most other species of sharks, it is an inhabitant of the open ocean in waters over 650ft (200m) deep and above 70 degrees F (21 degrees C). It is the most likely company at mid-ocean air or sea disasters and is therefore potentially dangerous to people.

During the Second World War, many naval tragedies saw sharks, mainly oceanic whitetips, in attendance. When a German submarine torpedoed the troopship *Nova Scotia* off the northern Natal coast of South Africa, there were only 192 survivors from a crew of 900. Many of the corpses had limbs missing, the result of shark attacks. Similarly, the USS *Indianapolis* was torpedoed by a Japanese submarine in the South Pacific. Of the 1,199 people who are thought to have made it off the stricken vessel, only 316 survived. Survivors tell how they could see down into the clear water as far as 25ft (7.6m), and would watch groups of 10ft (3.1m) long sharks circling below. Every so often a shark would dart up to the surface, grab a victim and tear off limbs or chunks of flesh. This bold behaviour seems to suggest that the predators were oceanic whitetips.

Jacques Yves Cousteau recognised the oceanic whitetip to be a dangerous shark. In fact, he considered it to be one of the most dangerous because of its apparent fearlessness. Unlike many other species which either circle a victim or approach it from below and behind, it will swim directly up to any object it considers potential prey and bump it while investigating its nutritional value.

Another species that could well turn up at a disaster is the blue shark (*Prionace glauca*). Blue sharks arrive in gangs, and will circle. Sometimes they will stay for a quarter of an hour or so without incident and depart peacefully, while at other times they will slowly, almost casually, tighten the circle before moving in for the attack, often inflicting a 'test bite' before the fully fledged assault. They seem to be intensely curious, an adaptation perhaps to their open ocean life where anything vaguely edible must be examined.

About 30 species of sharks share the same curiosity and have been implicated in shark attacks on people. Those that have had the occasional feast on human flesh include the great hammerhead (*Sphyrna mokarran*), shortfin mako (*Isurus oxyrhinchus*), porbeagle (*Lamna nasus*), grey nurse shark (*Eugomphodus taurus*), blacktip (*Carcharhinus limbatus*), Galapagos shark (*Carcharhinus galapagensis*), Caribbean reef shark (*Carchrhinus perezi*), and the grey reef shark (*Carcharhinus amblyrhynchos*).

One of the larger species capable of doing considerable damage to a human is the deep-water Greenland sleeper shark (*Somniosus microcephalus*), one of the few sharks to frequent polar seas. It feeds on seals during the summer months, and could easily mistake a person for a seal. Once a Greenland shark was caught and, when opened up, a seaman's boot was found in its stomach. Mats Heimersson, writing to the letters page of *National Geographic*, recalls a time when he was stationed at the Black Angel Mine, near the village of Umanaq in Greenland in the 1970s. From the local people he heard about two men who went to sea in a small boat, and they were caught in a storm. Both perished. Sometime later a Greenland shark was caught and, when it was opened up, the stomach was found to contain the remains of clothes belonging to one of the lost men. Whether the storm victims drowned before they were consumed, is not known.

The most bizarre shark attack, however, took place at Loch Lomond, Scotland . . . yes Loch Lomond! The victim was a 46-year-old American tourist and she encountered her shark in the Drover's Inn at Inverarnan. The shark – a great white – was a stuffed one hanging from the ceiling in the well-known Scottish hostelry. The incident happened when she jumped up to pat the shark on the nose and slashed her hand on the razor-sharp teeth.

# CHAPTER 12

---

# WHITE DEATH

Top man-eater is considered to be the great white shark, otherwise known as the white pointer, man-eater or white death. Found in both temperate and semi-tropical seas, this powerfully-built, torpedo-shaped shark grows to 20–23ft (6–7m) long, and mature specimens are quite capable of biting an adult human in two and swallowing the pieces whole.

One region where white shark attacks have become head-line news is the Pacific coast of North America. Here, great white sharks are active in the infamous 'Red Triangle', a stretch of California's Pacific coast bounded by Monterey in the south, Point Reyes in the north, and the Farallon Islands in the west. It is a mecca for great whites and they are not here primarily to attack and eat people, but to feed on seals, particularly elephant seals.

During the early 1970s the intensively hunted elephant seal was on the brink of extinction, so a ban was imposed on killing them. Gradually, their numbers recovered and they recolonised their traditional breeding sites. The conser-vationists were pleased . . . and so were the sharks. Great whites now congregate near breeding beaches and lie in wait, starting in early winter, the time when the bulls and yearlings arrive. Unlike the female seals that turn up somewhat later, the large bulls wait about in the water, unsure of their status on the beach. When they emerge they have to be ready to do battle to be beachmaster.

Bull elephant seals are big animals – up to 16ft (5m) long and weighing 5,000lb (2,270kg) – and so the sharks must be careful in choosing their targets and even more precise about the way they attack. Studies have shown that a shark approaches its target from below and behind, more commonly against a rocky background where the shark is less visible than, say, a sandy background. The attack itself is sudden, the shark relying on stealth. It accelerates into the target, using its weight and speed to take a large mouthful on the first pass. Its teeth – serrated, triangular meat-slicers in the upper jaw, and more pointed, grasping teeth in the lower – slice through skin, muscle and bone with ease.

California great whites (but not elsewhere in the world) then do something surprising – they hold back. It is as if they are waiting for the debilitated prey to weaken from loss of blood before going in to finish off the meal. After all, a bull elephant seal is a large, powerful and potentially dangerous foe. Its teeth and the claws on its flippers could inflict serious damage on a clumsy shark, so it seems to make sense to stand off and delay feeding until the prey cannot fight back. The first attack might also be an investigatory bite, just to check out the quality of the food. The great white has good eyesight, and uses this to home-in on a target. It is therefore able to spot potential prey from a distance (if the water is clear), but it cannot ascertain whether it is palatable. The first debilitating bite might serve to taste-test the food.

The problem for people in the water in the vicinity of a great white shark is that a surfer on a surf-board or a person in a kayak bears a striking resemblance to a seal at the surface, particularly when viewed from below and silhouetted against the sun. Could great white shark attacks on people, therefore, be cases of mistaken identity, and could the increase in attacks on people along the California coast be linked directly to the increase in the numbers of elephant seals? Scientists believe there is a connection. Since 1972, when the act preventing the killing of marine mammals was introduced, the frequency of great white attacks on people went up from one or two per year in the 1950s to five or six a year in the 1980s. There were over 50 attacks in the ten years between 1973 and 1983, 13 of which were surfers.

At Dangerous Reef off the coast of South Australia (where the real-life sequences for *Jaws* were filmed), John McCosker of the Steinhart Aquarium, Timothy Tricas (who has studied sharks off Australia and the USA), and underwater photographer Al Giddings carried out a series of rather macabre experiments using tailors' mannequins dressed in black wet suits. When placed upright on the sea-bed, the circling great whites showed little interest, and would only attack if the models were laced with bloody tuna steaks or other pieces of fish. If the models were strapped to a surfboard at the surface, however, the sharks reacted cautiously at first but then attacked.

Watching from above and below the surface, the observers were able to see that the shark relied primarily on surprise, approaching its victim – as previous research had suggested – from below and behind. Seen from the shark's point of view, the researchers could also see how a diver, surfer or bather could be mistaken for an aberrant seal. As part of their hunting strategy most predators have a 'search image' which enables them to quickly establish what is good to eat and where it is hiding. From below, a surfer on a surfboard has the right image. Predators are also programmed to seek out the sick and the disabled; after all, they are more easily caught and less likely to fight back. A human in the water must resemble a very 'sick' seal.

In addition, the great white shark's apparently catholic dietary habits, including humans encased in neoprene, may be due in part to the fact that it is swimming virtually blind during the last few centimetres before impact. The shark's eyes roll back in a socket to protect them from flailing claws. The snout is raised, the jaws protrude and the animal is guided on to its target by electrosensors in the snout. By the time the shark has realised its mistake and the expectation of a juicy seal steak turns into the reality of a wooden surfboard or a mouthful of foul-tasting wet suit, it is too late. The shark slams into its target, and programmed by a 'bite-and-wait' form of predatory behaviour which has taken millions of years of evolution to perfect, it waits nearby for the surfboard to 'die'.

Shark attack statistics appear to bear all this out. They show that a surfer or swimmer close to a seal rookery is more likely to be attacked than a surfer or swimmer elsewhere. These places are what scientists have called 'attack prone microsites', and they occur at many places along the California coast such as the Isla de Guadalupe off the Mexican coast, the Farallon Islands, Ano Nuevo Island, and Point Conception – all places where there have been numerous great white shark attacks. Also, people are more likely to be attacked if they are close to rocks than if they are on an open beach. The rule is not universal but it is generally the case.

The *Monterey County Herald* featured an attack, at 5.30 p.m. on 30 June 1995, that took place about 200 yards (183m) off Blue Fish Cove, in the Point Lobos State Reserve, just south of Carmel on California's Monterey Coast. An electrical engineer was diving in 90ft (27m) of open water with the aid of a small electric scooter, when he spotted a 'massive pectoral fin attached to the end of a torpedo-shaped body' about 20ft (6.1m) away, at the edge of his peripheral vision. At this time he was about 200ft (61m) from the Zodiac dive boat, in mid-water at a depth of 40ft (12m). He turned but the shark had already disappeared. Returning slowly to the surface, he looked round to see a tooth-filled mouth, about 2–3ft (0.6–0.9m) across, approaching rapidly. A second later he felt a dull pressure on his body as the shark's massive mouth clamped on to his thigh, torso and shoulder. Miraculously, the shark let go. A companion diver who had been slightly closer to the surface described the shark as moving 'like an express train going underneath me'. The victim reached hospital quickly and survived the ordeal. He suggests that his wounds were relatively slight because he had been sandwiched between the metal of his air-tanks and the DiveTracker (which he had invented) on his abdomen. Nevertheless he had puncture wounds about 30in (76cm) across, indicating a great white of about 16.4ft (5m) in length.

Significantly, the attack took place over rocky outcrops broken up by sandy channels – just the sort of sea-bed background associated with most great white attacks. There is also speculation whether the low frequency noise of the scooter attracted the shark, and the DC motor created an electrical field worthy of its investigation. The victim's diving partner revealed that seals were swimming in the area at the time, and has suggested that the shark was just investigating. If it had been a full-blown feeding attack, the target probably would not have spotted the shark before it hit.

Great white sharks are not fussy eaters, but there is evidence to suggest that they do 'taste' their victims before feeding properly. They prefer blubber or meat with a very high fat content and therefore, given the choice, would reject human flesh. The animal feeds infrequently and so must maximise its energy intake at every sitting. Seals and sea-lions (as well as dead whale carcasses on occasions) generally provide the right amount of blubber for the larger great whites. Even the grossest of people have insufficient fat to satisfy a great white. The result of the curious stand-off behaviour and the specific dietary preferences means that some people have survived attacks. If they are not too badly wounded, such as the severing of a major artery, and have not lost too much blood by applying a makeshift tourniquet, there is the slender chance of them remaining alive.

Such is the notoriety of this creature, however, that almost every attack on a person is headline news. Just looking at those incidents in the past couple of decades shows how the 'Red Triangle' has lived up to its name. In 1981 a surfer died when bitten almost in two when paddling his board out past the breakers on Moss Beach in Pebble Beach. The same year, a knee-boarder was killed in Spanish Bay, and on 19 December 1981, a surfer succumbed to a fatal attack from a great white at Monterey. A crescent-shaped bite was found in his abandoned board and later his body was discovered with much of the left side of his body above the hip missing. From the bite mark the attacker was estimated to be over 20ft (6.1m) long. A free diver was bitten but survived off Monastery Beach (next to Point Lobos Reserve) in 1985, and the following year, in December

1986, a great white estimated to be about 15ft (4.6m) long seized a diver off Monterey.

By September 1987, it was revealed that there were 27 attacks on swimmers and surfers (four fatal) along the California coast during the year. In June 1988, a great white estimated to be about 20ft (6.1m) long cruised in shallow water off the beach at Malibu, home to many international movie stars, before heading north to San Francisco. In February 1989, two students canoeing along the California coast, near Oxuard, Malibu, were killed by a great white. In 1990, a woman swimmer survived an attack on the leg, at Monastery Beach, and in September the same year, a Californian canoeist, paddling just 10 yards (9m) away from the shore, was knocked out of his craft by a great white, grabbed around the shoulders, and spat out. He got back into his canoe and paddled for the shore, lucky to be alive. In June 1995 a young female canoeist was attacked by a great white near San Diego. She survived with minor injuries.

The *Monterey County Herald* carried another report of an attack on a diver, accompanied by two friends, in Tomales Bay on 13 August 1996. The attack occurred at 11.40 in the morning and the victim was bitten on the torso, probably by a great white. The *San Francisco Examiner* of Sunday 6 October 1996 reported an attack by a large shark (probably a great white) on a surfer near Dillon Beach, north of Tomales Bay, California. The surf was good and many people were in the water when the shark struck at 9.30 on Saturday morning (5th). The surfer escaped with lacerations on the knee and hand. The attack followed another in the same area less than two months previously. Not far from Bird Rock, an abalone diver was bitten by what was thought to be a 15ft (4.6m) long great white. An underwater vehicle, similar to the one mentioned earlier, was being used during a similar incident – but this time a fatal one – that occurred off San Miguel Island, Santa Barbara, on 9 December 1996. On this occasion, an urchin collector who had been using a scooter was just handing the vehicle to a crew member when he was grabbed and killed.

Further to the north, Oregon shores have also played host to great whites. In October 1988 a surfer at Cannon Beach,

Oregon, needed 21 stitches in his knee after pulling his leg from the jaws of a great white. And, in April 1998, a 50-year-old surfer in the waters off Lincoln had a 15ft (4.6m) long great white slam into his board. It clamped the teeth of its upper jaw into the man's right thigh and those of its lower jaw into the surfboard. The man was shaken like a doll, pulled under twice, but then the shark simply spat him out. He punched the creature continuously around the dorsal fin and it dived. Unfortunately the surfer's ankle cord was caught in the shark's mouth and he went under again. The cord broke and he popped up to the surface. Significantly, the man said there had been many seals in the water before the attack.

Stinson Beach, California – another place where seal and sea-lions congregate – was the site of the next great white incursion. A 16-year-old boogie-boarder was about 50 yards (46m) off the beach when a great white, estimated by eye-witnesses to be about 10ft (3m) long bit into his thigh. The boy grabbed the shark's gills slits as he was pulled under and jerked hard until it let go. A foot-long gash in his thigh needed medical attention. The incident took place in August 1998. The reports go on, an increase in elephant seal populations paralleled by an increase in great white shark attacks on people.

Soon to take the title 'great-white-shark-capital-of-the-world', however, could be Chatham Island in the south-west Pacific Ocean. Here, a similar situation to that in California is developing. The local fur seal population is building up rapidly after legal protection, and sightings of great white sharks and attacks on people are increasing too. An attack on an abalone diver in 1996, for example, was the second in 12 months. Whether there are actually more great whites in the seal breeding areas or whether it is just that they have suddenly become evident is not clear. It is thought the slow breeding cycle of the great white would mean that sharks are unlikely to keep up with the current population explosion of seals. What is clear, however, is that the population of great whites is well fed and healthy.

Elsewhere in the world, wherever there are large seal breeding colonies there are sure to be great white sharks. And, wherever there are great whites, seals and people in the same waters, such as South Africa, Australia, Chile and in the Mediterranean, there are sure to be casualties. Outside of California, however, the bite-and-wait behaviour is not so evident. The sharks slam into their targets and eat immediately.

South Africa, particularly coastal waters – such as False Bay – near the Cape, are prime great white shark habitats. Here, there is an abundance of food, such as fur seals, sting-rays, and the youngsters of other species of sharks, but occasionally people are attacked mistakenly. On 11 April 1971, for example a 53-year-old German holiday-maker was about 100ft (30m) from a group of bathers on a sandbank beyond the surf line when he was confronted with a great white shark, estimated by eyewitnesses to be about 30ft (9m) long. It slammed into his stomach, ripping away the abdominal wall. He swam three strokes and then lay still, face down in the water. The shark repeatedly attacked the inert body, tearing off pieces in front of a beach filled with horrified bathers. The remains of the body were collected by a rescue craft and taken to nearby Knysna. A less disastrous yet nevertheless terrifying encounter took place later in the year. On 16 December 1971, a 16-year-old girl and some companions dived off some rocks near Jaeger Walk at Fish Hoek, False Bay, when she was grabbed by the wrist by a shark. It pulled her below but let go and as she bobbed to the surface and yelled, two lifesavers on surf skis heard her shouts and raced to the rescue. One of them fell into the water and the commotion frightened off the shark. They were able to get the girl to the shore and to medical help.

On 14 April 1974, a 19-year-old boy at East London also had a lucky escape. He was treading water beyond the breakers when he was bitten on the leg by a shark. He bunched himself up and punched at the fish for all he was worth. With a swish of its tail it swam away rapidly and the boy returned to the shore with minor injuries. Over a year later, on 17 August 1975, a 21-year-old was surfing at Cape St Francis when a shark bit into his thigh and lower leg but also took a chunk out of the surfboard. The boy was able to slide off on the opposite side and escape as the shark repeatedly attacked the board. A year after this incident, a 21-year-old Australian 'knee-board' surfer was attacked near Seal Point in Jeffries Bay. A companion saw the fin slicing through the water, but the warning was to no avail. The shark – a great white – slammed into the boy's left thigh, catching its lower teeth on the surfboard. The boy continually punched the shark, catching his finger between the teeth and having the flesh torn away. The shark pushed boy and board through the water for about 16ft (5m) and then swam off. On reaching the shore, he found a large flap of skin and muscle hanging from his thigh, but after a spell at Port Elizabeth Hospital he made a full recovery.

The following year, on 17 March 1977, a man was swimming in warm, turbid water near a river mouth at Still Bay. He was trying to cool off, and was body-surfing about 165ft (50m) out. As he was about to ride on a wave a shark grabbed his ankle and pulled him under. He kicked out with his other foot and the shark released him. He escaped with a damaged ankle, torn tendons and a severed nerve. After a few days in hospital he made a full recovery. Trennery's resort in the Trankei was the site of an attack on 19 December 1977. A 19-year-old boy was surfing about 985ft (300m) from shore, when a shark bumped his surf board and he was thrown into the water. At first the shark seemed to be preoccupied by the board, biting the skeg off, but then it took to circling the boy. It gripped both his legs and lifted him out of the water. Then, just as suddenly as it arrived, the shark simply swam away. The boy, however, lost one of his damaged legs.

In September 1987, a South African surfer on a Cape Town beach survived an attack by a great white shark that had grabbed both him and his surfboard in its mouth. In June 1990, a 21-year-old woman was killed by a great white at Mossel Bay, South Africa. July 1997 saw a 25-year-old man killed by a great white at Breezy Point on the Transkei Coast, south of Durban. The man was the first to enter the water and was alone as the tide rose and the surf began to swell. Friends on the shore could only see a surfboard and a pool of blood where the young surfer had been. Twenty minutes later his mangled body was washed ashore. The shark had taken his right leg. Two attacks on the same day took place in December 1997. In one incident, a surfer was bitten on the legs at St Francis Bay, 60 miles (97km) west of Port Elizabeth. The second occurred in Pringle Bay, about 44 miles (71km) south-east of Cape Town. A 39-year-old man was spearfishing in turbid water about 1,312ft (400m) from the shore. Witnesses, including the man's family, said they saw a 3ft (1m) tall dorsal fin and then there was just a pool of blood where the man had been standing. A search produced his speargun but failed to find his body.

During 1998, there was a spate of attacks on bathers, divers and boogie-boarders swimming off beaches along the South African coast. Various sharks, including great whites, were involved. Six attacks in five weeks was the tally along the Eastern Cape coast up until 23 June.

The South African body- (boogie-) board champion was the first to be hit at Keurbooms Beach, Plattenberg Bay in mid-May. He lost his right foot to a shark, and later had his right leg amputated above the knee. He also had serious injuries to his left leg. Then, two teenage surfers were attacked within hours of each other at Jeffrey's Bay and Sardinia Bay, near Port Elizabeth. The boy at Sardinia Bay was just 60 yards (55m) from the shore. The attack was a classic white shark: 'I didn't see the shark,' the 15-year-old boy said. 'It came at me from the bottom and knocked me off my board. I went under and there was a lot of thrashing about, and then I felt my arm go numb.' The shark had almost bitten off his arm. Nevertheless he used his surf board to shield himself, while a wave took him back to shore. Doctors later saved his arm.

On 30 May another bather was attacked at Pollock Beach, Port Elizabeth, and on 5 June a surfer was tossed from his board also at Jeffrey's Bay. Then, on 23 June, a 20-year-old body-boarder was attacked at Gonubie Point, near East London, by what was thought to be a great white. He was bitten on both hands, his right calf and left thigh, where an artery was severed. Witnesses said they saw the shark lift the young man out of the water, as he tried to punch it. Two men entered the water and pulled the victim out. The attack was so devastating that he had to have his left leg amputated, and his condition in hospital was critical and unstable some days later. Eventually he died of his injuries.

Elsewhere, the attacks continued. In early July, for example, a father wrestled with a shark that attacked his 10-year-old son. The boy had been boogie-boarding in just knee-deep water at Lookout Beach, Plettenberg Bay, when he was bitten on his right leg. His father reached down and grabbed it, but the shark bit him too. Eventually he wrestled it to the shore where it was killed. It was a 4.3ft (1.3m) long ragged-toothed (sand tiger) shark, a species not often implicated in shark attacks on people.

On 1 August, a spearfisherman was attacked at Pringle Bay (part of False Bay, near Cape Town), site of a fatal attack on a diver earlier in the year. On this second occasion, a great white shark estimated to be 16–18ft (4.9–5.5m) long grabbed the diver's leg just above the ankle. After a struggle, using his speargun, he was able to discourage the shark, and despite his severe injuries was able to swim the 66ft (20m) or so to the nearest rocks where his girlfriend hauled him ashore and ran for assistance. The crew of the rescue helicopter that took the young man to hospital said they could see the shark still circling just offshore.

During the same weekend, another attack took place at Fish Boma near Knysna, on Cape Town's south coast. A 19-year-old male surfer was grabbed by both legs by a 12ft (3.6m) long great white shark. The shark attacked in the familiar great white manner, from below and behind. Friends saw him disappear below the surface, but the boy was able to poke the shark in the eyes and it let him go. Two brave friends rushed to his rescue and managed to get him back to the shore. He survived.

Local sea temperatures were thought to be up on the seasonal average by about 6 degrees C (11 degrees F), but whether this has been significant remains to be seen. Local marine biologists have suggested that chumming for sharks around shark cages by shark-dive tour operators not far away at Dyer Island, where great whites congregate to take fur seals, has brought sharks to expect a free hand-out from people. If they fail to deliver they get chomped.

Hotel owners in the holiday resorts between East London and Saldanha Bay began to speculate whether a real-life *Jaws* scenario was developing. Was there a monster shark with a taste for human flesh hunting along this stretch of coast? Some beaches played safe – they closed. During 1999, the attacks continued. In January, a paddle-skier was attacked by a great white estimated to be about 13ft (4m) long. The man was about 50 yards (46m) off Bonza Beach in Eastern Cape province when the shark sunk its teeth into the paddle ski, leaving a 16in (42cm) bite mark. He punched the shark on the nose and it disappeared. In July 1999, a 14-year-old boy was surfing about 50 yards (46m) from the shore at Buffels Bay, near Cape Town and the Cape of Good Hope, when he was hit by a shark on his right side. A friend helped the boy reach the shore, but he was dead before he reached the hospital.

Most attacks by great white and other species of sharks in South Africa (mainly bull and ragged-tooth sharks) occur in the warmer water along the shores of the Indian Ocean, but there have been a few on the colder Atlantic coast. In 1920, a swimmer off Cape Town pier was attacked, and on 1 November 1942, at Fourth Beach, Clifton, an 18-year-old was killed when a large shark first took both his legs in one bite, and then returned for the rest of his body. The man was never seen again. Many years later the same beach was the site of another attack. The date was 27 November 1976. A 19-year-old boy and his friend swam about 820ft (250m) offshore and were treading water when one of them was hit by a shark. The boy 'felt a hard bump on my side' and 'a vice-like clamp on my chest', and he was carried through the water. The two shouted for help, and people in a dinghy came to help. The injured boy was taken to shore and eventually off to hospital. The other boy sought safety on a nearby yacht and was taken in by tender to the shore. Despite the description of a 'vice-like grip', examination of the injuries indicated that the shark had not bitten hard. It was as if it had been 'tasting' the victim, and had decided it was not to its liking. It also swam past the injured boy as if it was checking him out.

While South Africa has the doubtful distinction of being home to the largest number of dangerous sharks in the world, the island continent of Australia comes a close second. Great whites are found off Queensland, New South Wales, Victoria,

Tasmania, South and Western Australia. In March 1977 an incident occurred at Moreton Bay, 10 miles (16km) to the north of Brisbane. Three fishermen were thrown into the water after a freighter had sliced their boat in half. For 36 hours the men clung to a floating icebox, but just 45 minutes before their rescue the sharks arrived. At first small sharks bit at their bodies, but the blood and commotion attracted a 19.5ft (6m) long great white shark. Two of the fishermen were killed while the only survivor escaped by climbing into the icebox.

At Dangerous Reef, South Australia, three divers were surprised by a great white. The three were holding hands and the shark took the middle diver. The other two report hearing a noise like a lorry, thought to be the sound of the water entering the shark's mouth as it was opened for the attack. The victim was never found.

On 4 March 1985, a woman swimming in just 6ft (2m) of water at a public beach at Peake Bay, near Port Lincoln, became South Australia's sixth fatal victim of a great white shark attack. The shark was estimated to be about 20ft (6.1m) long, and it bit the woman in two. The attack was one of the few cases of a great white actually feeding on its victim. In March 1989, a surfer from Adelaide was savaged on the leg and bled to death. He became the seventh South Australian fatal shark attack victim.

In June 1995, a woman scuba diver was killed by a great white off the north coast of Tasmania. She was diving off a seal colony and was taken after having done a backward roll from the dive boat. Five days later, at Julian Rocks, Byron Bay, on the north coast of New South Wales, a scuba diver was attacked and killed after pushing his newly-wed wife out of danger at a dive safety stop. Other divers in the area at the time told of being circled by a great white. The shark was hooked by an angler not long after but it got away. It was estimated to be 20–21ft (6–6.5m) long. The next day, pilots of light aircraft reported seeing a large shark at the surface of the sea, 93 miles (150km) south of Julian Rocks, with what looked like buoyancy problems. An abalone diver was taken by a great white shark off Western Australia on 12 September 1995. Whales had been calving in the water nearby and there was blood and afterbirth in the water. The diver was hit at the surface, within minutes of entering the water.

On 27 October 1997, a great white attacked two surfers on a beach near Perth, Western Australia. Eyewitnesses estimated it to be about 18ft (5.5m) long. It disappeared, and was not seen again until 2 November when it attacked and ate a sea-lion in full view of spectators on the beach. The next day a lawyer was paddling a double fibreglass surf canoe with a friend in the surf on Cottlesloe Beach, 3 miles (4.8km) to the south of Perth, when he was attacked by what was estimated to be a 16ft 6in (5m) great white. The shark chewed the canoe in half, and the 51-year-old man suffered scratches to the face when the shark's tail hit him into the water. Another canoe picked up the man, and the shark was tracked and filmed by a rescue helicopter. Whether this was the same shark that featured in the previous report is not known.

In November 1997, a diver on an underwater scooter escaped injury from a 16ft (4.9m) long great white by ramming the shark on the nose and high-tailing it back to the dive boat. The incident took place near Albany, Western Australia. The shark lunged at the machine and bit the diver's arm. He needed 20 stitches.

On 30 June 1998, a great white shark killed a fisherman snorkelling at South Neptune Island, South Australia. The man was checking a net in shallow water when the shark struck. His companion managed to drag him from the water and on to the island, but he died from loss of blood. Why anyone should swim in these waters, where there are more great white sharks concentrated than anywhere else along the south coast is a mystery. Police investigated the death. The South Neptune Island lighthouse keeper identified a 12ft (3.7m) long great white that had been given the nickname 'Kong' as the shark responsible for the attack.

Middleton Beach, about 50 miles (81km) south of Adelaide, South Australia, was the site of an attack on a 21-year-old female surfer. The young woman was in chest-deep water, ready to ride back to shore, when she felt 'something knock her leg'. Then she felt a bite and something thrashing about next to her leg. She managed to reach the beach and was rushed to hospital where deep wounds to her leg were sewn back together. The attack took place in December 1998.

In February 1999, a 35-year-old surfer was paddling out at Scotts Head near Coffs Harbour in New South Wales, when a shark – probably a great white – attacked. Before he knew it, the man's arm was in the shark's mouth but he was able to struggle free. The serrated teeth had done their damage, however, and the arm had to be amputated later in hospital. In May 1999, police gave up the search for the body of a 22-year-old windsurfer at Hardwicke Bay near Minlaton on the west side of the Yorke Peninsula near Adelaide. They found a windsurfer harness, scratched sailboard and pieces of a badly shredded wet suit. They suspect the missing person was attacked and consumed by a great white shark

Another great white shark attack site is Chile, on the Pacific coast of South America. On 29 September 1963, what was thought to be a white shark attacked a diver at El Panul, Chile. And, on 5 January 1980, a shark attacked and killed another diver collecting shellfish in Pichidangui Bay, on the coast of central Chile. The culprit was thought to be a great white. On 15 December 1988, a third diver collecting shellfish near Valparaiso, Chile, was attacked by a great white shark estimated to be about 16ft (4.9m) long.

In the north-east Atlantic, great white sharks are not common, but they do occur. They were often seen following the whalers who brought back sperm whale carcasses to the Azores, and one of the largest great whites on record was harpooned there. Some of the other Atlantic islands also have been visited. In December 1979, off the northern coast of Madeira, south-east of the Azores, for example, a swimmer was killed by a great white shark, the first such attack in the island's history.

The largest congregation of great whites in the north-east Atlantic, however, is in the Mediterranean Sea. Indeed, it may come as a bit of a shock to European holidaymakers to know that some of the largest great white sharks in the world swim in the Mediterranean, a fact that has not gone unnoticed to the inhabitants of the region down the ages. A Campanian vase, dated 725BC, that was recovered from an excavation at Ischia

– a volcanic island near Naples – has a painting of a shark attack. The culprit depicted is thought to be a great white.

In 492BC, Herodotus described a creature referred to as the kete attacking people, and the poet Leonidas of Taranto describes an attack on a sponge diver called Tharsys, that ended with the unfortunate diver being bitten in two as he tried to get back in his boat. The crew, so it is written, buried his top half while the bottom portion was devoured by the shark.

Oppian, in the 2nd century AD, described sharks. In a charming English translation of 1722, the presence of the great white shark in the Mediterranean was noted in rhyming couplets:

> White sharks the Fisher's Curse, force on their Way
> And ominous Hyaena's size their prey.

The teeth, in particular drew Oppian's attention:

> Who see the Shark's capacious Jaws disclose
> A thousand tusks erect in flaming rows
> Despised the tusked Boar.

It was not until the 16th century, however, that attacks on people were written down in a more scientific way. Professor of Medicine, Guillaume Rondelet was aware that great white sharks took an unfortunate interest in people. He wrote: 'This fish eats others; it is very greedy; it devours men whole, as I know from experience: for between Nice and Marseilles, where Lamies are sometimes caught, in the stomach has been found a human body with full armour.' Rondelet considered the great white shark, rather than a whale, to be a contender for the biblical story of Jonah. A shark has the capacity to store food in the stomach for long periods without it breaking down, and it can also regurgitate its stomach contents at will, everting its stomach right out of the mouth. It is hard to imagine, however, how a person would not be ripped by the great white's razor-sharp teeth on the way into its mouth, and again on the way out.

Biblical references aside, recorded shark attacks have

featured in the Bay of Monaco, Genoa, the Israeli coast, Egypt and Greece. The identity of the shark is often unknown, although the great white is usually thought to be the culprit. The records go back in history. In 1758, for example, a 20ft (6.1m) long great white shark grabbed a sailor who had fallen from a French frigate anchored in the Mediterranean. The giant fish was harpooned, and was said to have weighed 3,924lb (1,780kg). Rondelet's specimen was even larger. It weighed over 4,000lb (1,814kg) and measured 22ft (6.7m) long. It was caught off the French coast near Aix in 1829. Inside was the headless body of a man encased – so the story goes – in a suit of armour!

In more recent times, there have been mercifully few great white shark attacks on people. There were just 18 recorded between 1863 and 1962, but the proportion of fatalities is much higher than elsewhere in the world. In 1909, a 15ft (4.5m) long female great white shark was caught in a fishing net off Capo San Croce, near Catania, on the south-east Sicilian coast. When its stomach contents were examined the pathologists discovered the remains of three partially clothed human bodies – an adult male and female and a young girl. It is thought a tidal wave triggered by the Messina earthquake may have swept the group into the sea where they drowned. The shark scavenged on their bodies.

Some attacks, however, appear in suspicious circumstances. Take for example the case of Zorka Prinz which appears in the International Shark Attack File. The *New York Evening Sun*, dated 30 August 1934, carried a story from Fiume, in the former Yugoslavia. The story told of a mother who dreamed her daughter would be attacked by a shark and so she tried to dissuade her from swimming far from the shore. The daughter, however, was a strong swimmer. Declaring that she did not believe in dreams, the girl swam towards a fishing boat in the sea off Reotore on the Italian side of the border in the Gulf of Trieste. One of the fishermen, according to the report, heard a girl scream, and set off in the direction of the sound, only to find a patch of blood-stained sea. The fisherman told reporters that he had seen a shark swimming around the edge of his net that day, and as far as the western press was concerned, the girl had been seized and eaten by the shark.

A few days later, on 1 September, a second article appeared. This time the dateline was from Belgrade and referred to the story of the girl and the shark. 'It is reported from Kraljevica in the Avala District', stated the article, 'that the news published by certain foreign papers according to which a young Yugoslavian girl, Miss Prinz, was attacked and eaten by a shark off the Italia coast, is without foundation.' The article went on to say that the girl was at her parents' home and she was about to take examinations for university entrance. Was the first newspaper story spurious, or did the community want to hush up the attack in order not to disrupt the tourist trade? Whatever the answer, another attack occurred a few days later. At Susak on 4 September 1934, an 18-year-old girl was fatally injured by a shark.

During the Second World War, there were many incidents involving ditched airmen, survivors from sunken ships and sharks. One such attack occurred in August 1943, when a US airforce pilot ditched in the Tyrrhenian Sea about 40 miles (64km) west of Naples. He was bitten on the legs and arms by sharks.

In July 1954 *The Times* of London reported that a Hungarian refugee and friend tried to swim from Pola (Pula) towards Fiume (Rijeka) but only one person made it, for the other was taken by a shark. July 1956 saw a great white shark attack on a British teacher in St Thomas Bay, Malta. The man did not survive.

In 1960, a diver off western Italy was attacked by a great white, but he lived and went on to catch his aggressor. In September 1961, a 19-year-old student was swimming with seven friends just 75 yards (69m) from the shore at Opatija, a seaside resort on the former Yugoslavian side of the Adriatic. He was grabbed on the left hand by a large shark which surfaced close to him. He lost the hand and his legs were mutilated, but the boy died before any boat could reach him. The same year, attention focused on Riccione, to the south of Rimini in the Adriatic Sea. Here, a skin-diver was luckier. He was trailing behind two companions when he was struck on the foot by what was identified as a great white shark. Two harpoons were released at the shark, which sloped off into deeper water.

On 2 September 1962, a scuba diver was spearfishing in the Tyrrenian Sea, off Monte Circeo, on the coast between Rome and Naples. He was descending to a depth of about 50–70ft (15–21m), when he was attacked by an unidentified shark. He died from loss of blood with multiple injuries to his legs. There was a bone-deep tear with loss of flesh from the lower thigh to the heel.

The Croatian coast has been the site of two unfortunate incidents. In September 1971, a Polish tourist swimming at Ika, near Opatija, was attacked and killed by a great white shark, and in August 1974, a great white killed a young German holidaymaker swimming off the coast at Omis, near Split. In 1976, a report emerged of an attack on a spearfisherman near Bizerta on the Tunisian coast by a great white shark.

In 1983 a shark scare in the Aegean islands of Greece ended when two large sharks were caught near to the shore and in the summer of 1985, there was a shark alert along a 62-mile (100km) stretch of the Italian coast, from Civitavecchia to Anzio.

On 1 August, the start of the Italian national holiday season, a women on an airbed was attacked by a 11.5ft (3.5m) great white in the Gulf of Genoa, and in the same year a 15ft (4.5m) specimen was caught and hauled out of the Golf of Lions.

One of the most controversial incidents occurred on Italy's Tuscany coast in 1989. On the afternoon of 2 February, a man was diving for lobsters in the Gulf of Baratti, near the port of Piombino, when an enormous shark, thought to be a great white, struck him and carried him away. Later his flippers, a still-fastened lead belt, and air bottles scarred with teeth marks were found. The newspaper headlines were predictable: 'Hunt for Killer Shark', Shark Fever Grips Italy', 'Il Terrore Degli Abissi'. Locally agitated groups of cafe proprietors, beach boys, and housewives discussed the pros and cons of great white shark biology. Everybody became an expert. Their fear was not of the shark, but a concern for their tourist trade.

Police scanned the depths with underwater cameras, while

army helicopters droned overhead. Crazed shark hunters set out floating buoys under which huge steel fish-hooks were baited with tempting chunks of mutton. Gawking spectators, armed with binoculars, came to witness the gruesome spectacle. Anything that resembled a shark, including dolphins and harmless sharks, was killed. One British tabloid carried a picture of a dolphin's dorsal fin, claiming it to be that of the killer shark. Our primeval fear of being eaten alive resurfaced. The shark became villain once more. It was the first reported shark attack on the Italian coast for 27 years, that is, if it really was genuine. A local magistrate revealed that the victim had been recently insured for £500,000 and traces of explosives were found on the recovered air bottle. The authorities had begun to suspect foul play. Nevertheless, shark fever continued unabated.

Interestingly, a 16ft (5m) long great white shark pitched up in the waters off Baratti, near Piombino, on 28 December 1998. The shark approached a fishing vessel 'in a non-aggressive manner', indicating that the area is certainly the haunt of at least one individual.

At about 3.30 in the afternoon of 6 June 1989, a surfer at the Tuscany resort of Marina di Carrara in north-west Italy was lying on his surfboard, waiting for a wave, when a great white shark estimated to be about 9.8ft (3m) long, first circled him and then suddenly slammed into him. He received wounds to his right thigh, but lived.

On 30 July 1991, a young man about 66ft (20m) offshore in a fibreglass surf canoe was attacked by a 11.5ft (3.5m) long great white shark in Tigullio Bay, Santa Margherita Ligure, near Genoa in north-west Italy. Tooth fragments embedded in the craft confirmed that the assailant was a great white.

Not all shark attacks in the Mediterranean, however, have been by large sharks. At 8 o'clock on the morning of 3 September 1993 a swimmer was about 656ft (200m) from the beach at Playa de les Arenes, Valencia, on the Spanish Costa Blanca when he was attacked by a small 6ft (2m) long shark. The shark bit off his toes. Whether is was a baby great white or another species of shark we shall never know.

Great white sharks – big ones, that is – also appear in parts of the world where attacks would be unexpected, particularly in temperate waters. The most northerly attack along the US east coast, for example, occurred in Buzzard's Bay, Massachusetts in July 1936. A 16-year-old boy was swimming in water only 10ft (3m) deep, just 150 yards (137m) from the shore. Suddenly, without any warning and with no sound, he was grabbed by the left foot and dragged below the surface. He fought his attacker, but grazed himself on its skin. It was a shark, thought to be a great white shark. Just 10ft (3m) away, his swimming companion Walter Stiles grabbed him and held him up. The shark – a large one according to Stiles – swam around the patch of blood spreading away from the two swimmers. Stiles managed to get the boy to the shore but after his leg was amputated about five hours later, he died in hospital.

# CHAPTER 13

---

# WATER-BEASTS

While sharks have dominated the headlines as the sea's primary man-eaters, there are other underwater predators that have featured in accounts of attacks on people. One that has received an undeserved bad press is the killer whale or orca (*Orcinus orca*). The Romans, among others, were responsible for setting the tone, describing the orca as *tyrannus balaenarum* – the 'tyrant whale'. In German it is the mordwal or murder whale.

Down the years, killer whales have somehow acquired a reputation for attacking and devouring people. Even the US Navy awarded it the highest danger rating in its 1960s *Diving Manual*. It stated: 'The killer whale has a reputation of being a ruthless and ferocious beast', and went on to suggest that 'if a killer whale is seen in the area, the diver should get out of the water immediately'. The US Navy's *Antarctic Sailing Direction* states that killer whales 'will attack human beings at every opportunity'. And, in one diving manual the author writes 'There is no remedy against an attack by a killer whale except Reincarnation'. The evidence for the concern, though, is slim.

The orca is the largest and fastest member of the dolphin family. Males – identified by their tall sword-like dorsal fin –

can be up to 23ft (7m) long and weigh 9,920lb (4,500kg), but it is the smaller females that dominate orca pods. Orca society is a matriarchy, with the oldest female leading the pod.

There appear to be at least two types of orca pods – the so-called resident pods that patrol recognisable home ranges close to the shore and feed mainly on fish, and the transient pods that not only live offshore and feed mainly on sea mammals, such as whales and dolphins, but also appear near to shore seasonally at the breeding rookeries of seals and sea-lions. The latter group could be potentially dangerous because, like great white sharks, they might mistake a person in the water or on a surfboard for an aberrant seal. Cases of mistaken identity have occurred in the past.

One of the most famous accounts is that of Lieutenant Henry 'Birdie' Bowers, who was with Robert Falcon Scott on his ill-fated expedition to the South Pole in 1911. On one eventful morning Bowers awoke to find that he and two other men, three ponies, and a pile of equipment were adrift on a section of ice. Several killer whales surfaced close by. Bowers wrote that 'their huge black-and-yellow heads with sickening pig eyes were only a few feet from us'. Despite their potentially dangerous predicament, people and horses were able to walk off the ice-floe unharmed.

On another occasion, the expedition's photographer Herbert Ponting was standing on a relatively thin ice-floe when a group of orcas crashed through the ice and momentarily set him adrift. Pontin was terrified but survived the ordeal. The whales had been curious about the seal-shape they had spotted through the ice, but were clearly dis-appointed when they found the 'prey' to be a person.

There are, however, the inevitable stories of people who did not survive. One was told by orca-watcher Paul Spong and included in *Mind in the Waters*. It tells of two loggers working in British Columbia. The year was 1956. The men were moving logs from a hillside into the water and noticed a pod of orcas passing below. One of the men deliberately let one of the logs go and it slid down the hill straight into the back of one of the whales. The whale was not seriously hurt, and the pod swam away. In the evening, the loggers were rowing back to camp when the orcas reappeared. They headed straight for the boat and turned it over, tipping the men into the water. One man – the one who had let slip the logs – was never seen again. The other lived to tell the tale. Unfortunately, there is no official documentation to support the story, and so it is probably no more than a loggers' tale, a modern moral fable.

There are also some stories that may have an element of truth, but they are more probably cases of mistaken identity rather than deliberate feeding attacks on people. Several yachts have been holed, for example, and orcas have been blamed. The most famous took place on 15 June 1972, when Dougal Robertson's 43-foot ketch *Lucette* was rammed about 150 miles (241km) west of the Galapagos Islands in the Pacific Ocean. Robertson's two sons spotted 20 orcas in the area before the yacht sank. An account of their 37 days adrift in a small life raft is recalled in *Survive the Savage Sea*.

On 9 March 1976, the Italian yacht *Guia III* was off Dakar and making for the Cape Verde Islands when it was holed. It sank in a few minutes. As the crew clambered into the life raft they noticed a pod of orcas nearby. The whales came close and then swam away. In both cases, the crews were not harmed, yet they were certainly vulnerable to attack. And it is by no means certain that the whales were the culprits for ramming the boats anyway. If they were, could it be that the orcas had mistaken the large dark shapes of the boats for baleen whales and had attacked without warning so as to surprise their 'prey'?

That orcas have little or no culinary interest in people was illustrated by an incident on 9 September 1972 at Point Sur, near Monterey, California. It is an area where many attacks on people by great white sharks have been recorded, but the creature that came up behind Hans Kretscher that day was something else altogether. Wearing a wet suit, Kretscher was lying on his surfboard about 100ft (31m) offshore when he felt something shove him from behind. He turned to see a glossy black shape, and thought it was a shark. The creature grabbed him and Kretscher banged it on the head with his fist. It let go and he was able to body-surf to the beach. The Pacific Grove Marine Rescue Patrol whisked him off to hospital where a hundred or so stitches were needed to close three deep gashes in his thigh. The surgeon, who had sewn up many shark-attack wounds, said that Kretscher's wounds were definitely consistent with the kind of slashes he would expect from an orca's teeth. Several witnesses at the beach consistently described an animal with a tall dorsal fin, white under-markings and a black back.

Significantly, seals had been seen in the area before the attack, and transient pods specialise in taking seals. Had the orca which attacked Kretscher mistaken him for a seal and having realised its mistake, moved on in search of more juicy prey?

There is no doubt that orcas are curious about humans, and will often come towards people and boats just to check them out. A predator must always explore any opportunity to feed, and in the wild, this is usually all they do – investigate but not attack. This was the case when Terry Anderson spent the night lashed to his wrecked trimaran off Baja California in March 1972. Orcas brushed close to the drifting boat but none attacked.

Scuba divers off New Zealand have had many encounters in the water with orcas. Wade Doak gathered several stories for an article in *Diver* magazine. One underwater photographer heard the high-pitched whining sound of his twin strobe lights recharging and then heard something mimicking the sound. Turning round, he came face to face with four large orcas, and he just stared back at them in awe. The orcas looked at first, then swam away.

Another diver was returning to his boat after looking for lobsters when he felt his foot hit something solid. It was an orca, and it took hold of one of his flippers. As he looked down, it let go and swam away. A few seconds later it was back, and then began to inspect him, swimming in from all angles until his boat arrived and pulled him from the water. Several other divers have had similar experiences, the orcas appearing to be more inquisitive than dangerous. On one occasion, a marine biologist from the Leigh Marine Laboratory was in the water when a 16ft (5m) long orca approached him. To be on the safe side, the man dived towards the sea-bed, but the orca followed. It then opened its mouth and put its jaws around the diver's right ankle, foot and flipper, but did not bite. The man extricated his leg and tried to swim away, but the orca took his left foot into its mouth. Still it did not bite. He then kicked the orca with his free foot, but each time the whale came back and continued his game. After the sixth mouthing, the diver thought enough was enough and dived quickly to the bottom. The orca simply disappeared.

In captivity, however, there is at least one case on record of orcas killing one of their trainers. The attack occurred on 20 February 1991 at an aquarium in Victoria, Canada. Three orcas – one male and two females – attacked a female trainer when she fell into their enclosure. One grabbed her in his mouth and dragged her round the pool. She swam free and was about to pull herself out of the pool when she was dragged below again; and all this in front of a horrified audience. The other trainers tried to distract the whales with fish, verbal commands for tricks and banging on the sides of the pool with buckets, but the orcas ignored the bedlam. As the girl came up screaming one last time, they took her to the bottom and drowned her.

The case was thought to be not one of man-eating as such, although the suggestion was made later that the animals were kept 'hungry' if they did not co-operate and perform their tricks. The alternative proposal was that it was a case of aberrant behaviour. Both females were pregnant and particularly aggressive to each other and the male. Indeed the male orca, named Tilikum – meaning 'friend' in the language of north-west American Indians – was driven by the females into a small veterinary holding tank.

Tilikum was to feature in another bizarre episode in July 1999. A man – a young drifter or tramp – was found lying naked on his back in its tank at its new home in Florida. His body was unharmed apart from a few scratches and bruises, and it is assumed he jumped, fell or was pulled into the tank and drowned. The whale probably played with him as he would a toy.

That orcas in captivity behave aggressively towards people is not a novel suggestion. Jay C. Sweeney, a veterinarian who contributed to the CRC *Handbook of Marine Mammal Medicine*, noted that 'aggressive manifestations toward trainers have included butting, biting, grabbing, dunking, and holding trainers on the bottom of pools and preventing their escape'. Whether this behaviour is simply a rather robust orca 'playtime' or whether the animals see us as potential prey is unclear. It may just be one way of relieving boredom. Orcas in captivity apparently get lethargic, neurotic and even become dangerous. Orca-watcher Eric Hoyt tells of an old bull he saw at one of the California aquariums that was kept in a pool away from the public. It had chomped on several dolphins and had made a grab for a woman. A reward of $500 was offered for anybody who dared swim across his tank.

In March 1987, a 21-year-old trainer was grabbed by an orca in front of many thousands of eager spectators. He was pulled to the bottom of the tank and then brought bleeding to the surface and released. As he waved stoically to the crowd, another orca slammed into him and dragged him to the bottom of the pool where it tried to drown him. When he finally escaped he had, according to reports, a badly cut body, a ruptured kidney, and a lacerated liver. Several other 'accidents' followed. An orca landed on a female trainer during rehearsals, a young man was rammed, a trainer was bitten on the hand while feeding, and a male trainer who was riding a female orca was crushed by a male orca. The man had fractures to hips, pelvis, ribs and legs. He was put back together again in hospital but has never fully recovered. The orca responsible was not new to such incidents. In 1978, he had pinned a female trainer to the bottom of his pool.

There are reports of a girl being bitten on the leg by a resident orca, and several trainers in other establishments who have been riding orcas and also have been held under and nearly drowned. Footage from home video cameras used by members of the audiences provides the evidence. In June 1999, a trainer was nipped by a 23-year-old female orca during a show. The man escaped injury by jumping quickly out of the water.

The orca's larger cousin, the sperm whale (*Physeter macrocephalus*) – the largest and most powerful hunting predator in the sea – is not averse to swallowing the occasional person, albeit under provocation. Extraordinarily, at least one person has survived such an ordeal. In February 1891, according to a report in the *Great Yarmouth Mercury* of October 1891, the crew of the whaler *Star of the East* were attempting to take a large bull sperm whale. The whale had smashed two boats and a sailor was lost overboard, presumably swallowed by the whale. The boat followed the whale, waiting for it to die of its injuries. When it did so, they brought it alongside and hoisted the stomach on board. Inside, doubled up and unconscious, was the seaman. When revived he first behaved 'like a raving lunatic' but with careful counselling by the officers and crew, the man's behaviour eventually returned to something resembling normality. Patches of his skin were bleached white where the whale's gastric juices had begun to act. Several notable scholars investigated the story and came to the conclusion that it was true, giving some credibility to the biblical story of Jonah.

At the other end of the world, on the harp seal hunting grounds off Newfoundland, a man was not so lucky. The year was 1893, but the story, curiously, was not to surface until 1947. Writing in the magazine *Natural History*, surgeon Egerton Davis recalled a seal hunter falling off an ice floe and being taken by a sperm whale. An accompanying schooner, the *Toulinguet*, fired a cannon shot and hit the whale. It swam away but was seen the following day, floating at the surface belly-up. The stomach was lifted on to the deck and cut open. Inside was the man, his chest crushed and his body surrounded by the lining of the whale's stomach. The exposed parts were partly digested.

Sperm whales have also been held responsible for attacks on boats. Whether the aggression was another case of mistaken identity or a deliberate attack is not at all clear, but at 8 o'clock in the morning the crew of a fishing boat pulling up fish traps off one of the Canary Islands was to discover that such an encounter has an inevitable outcome. The incident occurred on 16 September 1999, and the 33ft (10m) long boat was stationary about 492ft (150m) from the coast of Los Cristianos, Southern Tenerife. The depth of the water was about 394ft

(120m). As they were going about their business, the crew members were knocked over as the whale hit and lifted the bow. Within five minutes the boat was filled with water and the crew jumped into the sea. They were not attacked by the whale, thought to be one of the resident females.

Besides sharks, orcas and sperm whales, the marine creature that seems to evoke the most fear in people in the water and which appears to have an occasional appetite for human flesh is the giant barracuda (*Sphyraena barracuda*).

The giant barracuda is a thin, pike-like fish with a silver and black striped body that can grow up to 6ft (1.8m) long. Individuals have been known to enter shallow water and sever the leg of a victim paddling in the shallows. Some people consider it to be the 'most formidable bony fish in the sea'. Reports of it eating people, however, are mercifully few. It might get its evil reputation from its insatiable curiosity. The solitary giant barracuda and schools of smaller species have been known to follow swimmers, divers, people in boats and even people walking along the shore. They tend to use sight rather than smell to home-in on prey, and are attracted to shiny, flashing objects in the water.

One of the most bizarre attacks occurred in February 1999 when a holidaymaker and his wife were on a speedboat outing off the Caribbean island of St Lucia. The man was lying back with his arms above his head when he felt a sudden jolt. A 6ft (1.8m) long barracuda had leaped out of the water and sunk its 150 needle-sharp teeth into his arm. He was left with a gaping wound that required 40 stitches to close up.

Another voracious predator is the 2ft (60cm) long bluefish (*Pomatomus saltatrix*). Dubbed 'the most ferocious saltwater fish in the world', it is a fast-swimming predator (speeds of up to 30ft or 9m per second have been recorded) that gathers in enormous shoals. Someone once described it as 'an animated chopping machine', but does it live up to this savage reputation? When a shoal attacks mackerel, menhadens, striped bass, tomcod or herring – so the stories tell – the feeding frenzy leaves behind a trail of blood and body parts for ten times as many fish are torn apart as are consumed. Some say the fish is so voracious that it will feed to bursting point – up to 40 fish in one sitting, regurgitate its gut contents and start all over again.

Scientists, like those at the Sandy Hook Laboratory of the US National Marine Fisheries Service, might disagree. They look at the bluefish as a highly efficient predator, and consider that regurgitation of their food is more a response to being caught. Fishermen, however, are very wary of its razor-sharp teeth, and many have had chunks taken out of a carelessly placed hand.

Many commentators have wondered what would happen if someone should fall into a shoal of feeding bluefish. The conclusion has been that they would be reduced to a skeleton in a very short time! This is what happened to three luckless seabirds off Eastern Egg Rock, Muscongus Bay, Maine on 11 August 1975. The birds – a newly fledged guillemot and two common eider drakes – were tugged repeatedly under the water and were unable to fly away. They were being attacked by three fish thought to be bluefish.

The bluefish is found in most warm seas (except the west coasts of North and South America and Europe), including the Mediterranean, and appears off the US east coast where it migrates north in summer and south in winter.

Bluefish are caught by sports fishermen off the east coast of the USA, but occasionally they have turned the tables. In April 1976, for example, a ravenous shoal of bluefish terrorised Florida's Gold Coast and chunks were taken out of a dozen or so people unfortunate enough to be in the water as it passed by. A 17-year-old surfer was bitten on the foot at Haulover Beach, North Miami, and had to have 60 stitches. Another boy was seen to emerge from the surf with a bluefish clamped tightly to his hand. It is likely that the bluefish were chasing bait fish into the shallows and the bathers were caught up in the subsequent mêlée. A wriggling finger or toe looks much like a bait fish to an actively feeding bluefish. Another case of mistaken identity.

In December 1997, the beaches around Hollywood, Florida, were closed when an eight-year-old Canadian boy was attacked by an unknown sea creature. The boy was in 4ft (1.2m) deep surf when he received a slashing bite that left a gash from his left knee to his ankle. A shark was suspected, although bluefish and barracuda were also considered likely culprits, according to local experts.

Off the Pacific coast of Central and South America it is squid – giant squid – rather than fish that local fishermen fear. The Humboldt Current squid (*Dosidicus* or *Ommastrephes gigas*) is dreaded for its ferocity – it is considered a demon. It grows up to 12ft (3.7m) long and hunts in large shoals at night. Michael Lerner, benefactor of the Lerner Laboratory in Florida, once caught a specimen which, it was claimed, weighed 300lb (136kg) and had a body 10ft (3.01m) long and tentacles stretching to 35ft (10.7m). The eyes were 16in (41cm) across.

Sports fishermen, who have found that the squid often strip their prize marlin or swordfish to the bone, have turned to catching the squid itself. It is a curious sport. The fishermen go out at night and wear pillow cases over their head for protection against the squid's ink. When the squid is brought to the surface it not only shoots about at high speed, but also squirts a high pressure jet of water and ink which deluges the boat and the fishermen lining the rails.

Many of the hooked fish fail to reach the surface: the other squid eat them alive. The steel traces on fishing lines may be bitten in two by the squid's sharp, chitinous beak – shaped somewhat like a parrot's beak. The fate of any fisherman who fell over the side would not be hard to imagine. Indeed, in October 1998 reports came in of a fisherman who had fallen overboard being killed and eaten by a school of squid off Loreto, Baja California. Fishermen generally know that squid are nearby by the smell of ammonia, and the appearance of saucer-sized luminescent eyes at the surface.

In the late 1990s, a shift in the ocean currents of the Pacific – the phenomenon known as El Nino – caused warm waters to flush the California coast; and with it came the jumbo-sized squid. Wednesday was designated by one fishermen out of Bodega Bay as squid-fishing day. On his fish finder he could see the squid hovering in mid-waters during the day, rather than in deep water, as was once thought. As soon as one is hooked and hauled to the surface, however, it changes colour immediately and the rest of the school is whipped up into a frenzy.

Underwater wildlife film-makers, Howard Hall, Bob Cranston, Mark Conlin and Alex Kerstitch were once in the middle of such a frenzy. They were diving at night in the Sea of Cortez near La Paz. Their boat drifted over an area where the bottom was over 1,000ft (305m) deep. They had bait over the side and lit up the water with enormous 1,200 watt lights. The squid arrived in ones and twos at first, and Kerstitch went in to look at them and take some still photographs. Below him were groups of large squid flashing colours at each other in an extraordinary luminescent body language. In the water, he was immediately 'mugged' by three squid – each about 5ft (1.5m) long. At first he felt a tug, as a large squid wrapped its tentacles around his swim fin. It began to pull him down but as he kicked himself free another swam in and grabbed him by the back of the neck – the only part of his body not covered by his diving suit. 'I felt the cold embrace of tentacles with their sharp-toothed suction cups digging into my bare skin,' Kerstitch told reporters. 'It was like somebody was throwing a cactus on to my neck.'

Kerstitch was able to hit the squid with his diving light and it began to release its grip, but not before it grabbed the light and a gold chain around his neck. He made for the surface, but another squid wrapped its tentacles around his face and chest and dragged him under. Kerstitch dug his fingers into its body and it slid down to his waist, but it was also pulling him below. As he grabbed for the diving ladder, the squid made off with his decompression gauge, and Kerstitch was able to clamber aboard. He had a line of round, red scars circling his neck. This species of squid has serrated suckers which help capture and tear prey before it is sliced by the beak at the centre of the arms. Kerstitch had a lucky escape, although the sortie could have ended in tragedy. The three squid had pulled him down very deep before they let him go.

Some consider the squid among the most dangerous animals in the sea. This was evident during the Second World War. On 25 March 1941, the British troopship *Britannia* was sunk by the German raider *Santa Cruz* in mid-Atlantic. She was about 1,200 miles (1,931km) west of Freetown. Some men survived and clung to floating spars and life-rafts. Lieutenant R. E. G. Cox and eleven other men were supported by a very small raft; only their head and shoulders were above water. Sharks were the number one fear, but one night something far more sinister came up from the depths. A giant squid surfaced and wrapped a tentacle around one sailor and pulled him under. A little later Lieutenant Cox was seized by the leg, but the beast let go. Cox recalls incredible pain as the suckers were pulled off. The next day he noticed large ulcers had appeared where the suckers had gripped, and when he was rescued the medical orderlies were constantly treating the wounds. He wrote to Frank Lane, author of *Kingdom of the Octopus*, in 1956, telling him that the red marks were still there, and Professor John Cloudesly-Thompson at Birkbeck College of the University of London was allowed to examine them. He was able to confirm that they were sucker scars likely to have been from an attack by a giant squid.

While several fanciful reports in tabloid newspapers tell of monster squid rising to the surface and plucking people from sailing yachts – such as the creature that grabbed a woman and her daughter from a 50ft (15m) long ketch as it sailed over the Yap Trench in the Pacific Ocean – the *Britannia* incident is the only substantiated case of a giant squid attacking and killing people.

Giant squid (*Architeuthis*) grow to a staggering size, most of the length in the pair of retractable tentacles that are used to capture prey. The largest one reliably measured in recent years was a 47ft (14.3m) long individual found in 1966. It was injured after a fight with a sperm whale in the Tongue of the Ocean on the Great Bahamas Bank and picked up by a US coastguard ship. Larger squid have been claimed. In 1878, at Thimble Tickle Bay, Newfoundland, a giant squid was caught that had a body 20ft (6.1m) long and tentacles 35ft (10.7m) long. Giants are still found occasionally, making the giant squid the largest invertebrate (animal without backbone) alive today. Fortunately, it spends most of its time at great depths in the gloomy abyss. Its relative the giant octopus, however, is found closer to shore.

The largest known octopus is the Pacific giant octopus

(*Octopus apollyon* or *dofleini*) that inhabits the rocky Pacific coast of Asia and North America, from California northwards to Alaska in the eastern Pacific, and from Kamchatka to Japan on the western side. Mature males weigh 50lb (23kg) on average, and females 33lb (15kg). Average arm span is about 8ft (2.5m). The official record size is held by an individual that was gently wrestled to the surface (by rocking the beast from side to side) by skin-diver Donald Hagen in 1973. It had a radial span from arm tip to arm tip of about 23ft (7m). But there have been larger specimens claimed. In 1957, an octopus captured on the coast of western Canada was estimated to weigh nearly 600lb (272kg) and have an arm span of 31ft (9.6m).

The giant octopus is large and powerful, and will grip tenaciously with its suckers, but it tires quickly. This is because it has a copper-based blood pigment – haemocyanin – which is less efficient at binding with oxygen, as is the iron-based pigment haemoglobin in our own blood. Nevertheless, there are a few stories of it attacking and attempting to devour people. In October 1877, the *Weekly Oregonian* reported a North American woman going to the sea to bathe and being caught by an octopus and held under until she drowned. Her body was discovered, still in the monster's embrace, and some members of her tribe had to dive down and sever its arms before the corpse could be recovered.

In April 1935, a large Pacific octopus attacked a fisherman wading in waist-deep water near the entrance to San Francisco Bay. It wrapped its tentacles around the man's body, legs and left arm. He was being dragged to deeper water and cried out for help. Another fisherman ran to the rescue, and after a formidable struggle during which the rescuer hacked off some of the creature's arms, it was finally dispatched with a knife-blade between the eyes. It had a radial span of 15ft (4.5m) and weighed about 43lb (19.5kg).

Some years later Hollywood cameraman John Craig learned to have a healthy respect for the Pacific giant octopus. He wrote in his book *Danger Is My Business* about a time when he was lowered to the sea-bed, at San Benito Islands off Baja California, in an old-fashioned suit with brass helmet and lead boots. He spotted a large hole in the rocks, and entered a vertical shaft. At the bottom he found two large octopuses. Instead of rapidly back-tracking, he remembered the advice given to him by Japanese divers. They had said that you should not make any rapid movement that might attract the octopus's attention. Rather, they suggested, you should remain quite still. The octopus might examine you with its arm but will ignore you if you are motionless. Craig stayed absolutely still and, sure enough, the octopuses behaved as the Japanese had said. One raised its arm, examined the outside of the diving suit, and then seemed to lose interest. Then Craig made his move. He slipped off his lead boots, inflated his diving suit, and began to float to the surface. But the move was premature. The larger of the two octopuses stretched out an arm and grabbed his leg. Fortunately it was sitting on gravel and had no hold on the rocks; it was jerked out of the hole and hauled to the surface still attached to Craig. On reaching the surface, the diving attendants hoisted Craig out of the water and saw that the octopus had smothered him completely. They tried to tear it off and went about attacking it with axes. They kept one of the severed arms. It was 8ft (2.4m) long. This meant that the octopus was about 18ft (5.5m) across. If it had been on rock rather than shingle, it would have killed the diver. Whether it would have eaten him too is unclear. The octopus, like the squid, has a formidable beak well capable of slicing through a diving suit and human flesh but more usually used for breaking into the carapace (shells) of crabs.

Down the years there have been several similar incidents recorded. Some have been in the Mediterranean Sea, where the common octopus (*Octopus vulgaris*) once grew to a large size. In his *Excursions in the Mediterranean Sea*, Sir Grenville Temple wrote about a Sardinian sea captain who was bathing in just 4ft (1.2m) of water when he was seized by an octopus and held under. When his body was discovered the octopus had twined its arms around his limbs.

Another sinister tale is told in Clemens Laming's book *The French in Algiers*. He tells of soldiers going to bathe in the sea each evening. From time to time, however, some did not come back. As a consequence, bathing was strictly forbidden. Nevertheless, several men went to bathe one evening. As they swam up and down the rocky coast, one of them suddenly let out a scream for help. The others rushed to help him and discovered an octopus had caught him by the leg. Four of its arms held the man, while its other four were attached firmly to the rock. Outnumbered, the men eventually killed the creature and brought it home. Out of revenge they boiled it alive and then ate it.

In August 1867, the *Genoa Gazette* had news of a carter who went to bathe near the reef at San Andria and was seized by a large octopus. There were many other bathers in the area who witnessed the event but none was brave enough to go and help the struggling man and he was drowned.

Admiral Baillie Hamilton once wrote to Henry Lee, naturalist at the Brighton Aquarium in the late 1800s, telling of a particularly large octopus that lived in Gibraltar Bay. It had grabbed and drowned one of the soldiers who guarded the Rock.

In the Far East and south-west Pacific, coastal peoples also have stories of dangerous octopuses. The South Sea Islanders, for example, are fearful of these creatures and, if the stories are to be believed, they have good reason to be wary of them. In September 1984, two fishermen from the Pacific nation of Kiribati, who had been hunting octopuses with spears, were reported to have been held under water and drowned by several octopuses that were estimated to be about 12ft (3.5m) across.

Even more ominous were some very matter-of-fact references to giant octopuses in a Japanese book in three volumes called *Land and Sea Products* and written by Ki Kone. The book is about Japanese fishing methods and fish-curing, but it contains some interesting illustrations which appeared in *The Field* of 14 March 1874. One picture shows a fisherman in a small boat being attacked by a giant octopus. The fisherman has reached forward from the stern and with a long-handled knife, rather like a whaler's flensing knife, has chopped off one of the creature's arms. The picture does not seem to show any exaggeration: the man, the boat and the knife all seem to be drawn to the same scale. The octopus is huge. A second picture is of a fish market in which are hanging two cephalopod arms. The onlookers appear astonished at the size of the limbs. And, in a further illustration a fisherman is depicted catching cephalopods. He tosses crabs into the water and spears the unfortunate octopus when it rises to the surface to catch the bait. Some people have interpreted the pictures as fishing for giant squid, but the animal with the fishing boats looks very much like a giant octopus. And in the last picture, the fact that the fisherman is using crabs – an octopus's favourite food – suggests again that he is fishing for octopuses – giant ones!

Neil McDaniel, writing in *Oceans* in 1989, mentions 'rumours of commercial divers having spotted one hulking monster some 50ft in diameter in deep water off the coast of Japan'. Could the Japanese illustrations then be based on reality? A newspaper report about an incident off Mantiaco, in the southern Philippines, on Christmas Eve 1989, suggests they could. On that day, a motorised catamaran-style canoe was overturned by what one survivor described as 'an octopus the size of a cow'. Twelve people survived the accident by hanging on to the upturned hull, but one small boy was drowned. After its attack on the boat, the octopus appeared to ignore the people in the water and returned to the depths.

Other books have featured mysterious octopus tales. In *Great World Mysteries*, by Eric Frank Russell, there is a description by a diver of a huge black mass that drifted up from the depths in the South Pacific and engulfed a large shark the man had been following. And, in *Myths and Superstitions*, Julio Vicuna Cifuentes tells of stories from Chilean fishermen about a creatures they call 'the hide'. It is a large flat octopus that 'has the dimensions and appearance of a cowhide stretched out flat. Its edges are furnished with numberless eyes, and, in that part which seems to be its head, it has four more eyes of larger size'. The creature is said to take anybody who enters the water.

One of the most bizarre man-eating tales, however, must be that of the oil-rig diver who was swallowed whole by a giant grouper . . . lead boots, helmet and all! And if that seems an exaggeration, diver J. T. L. Ponds can assure us that it is not. He and a colleague (Jake) were diving on an oil-rig in the Gulf of Mexico and were being harassed by a particularly large grouper; it was big enough 'to swallow a Volkswagen', according to Ponds. Every time it opened its mouth a strong current would draw in fish and as it approached the divers they could feel the movement of the water. 'Suddenly, I seen Jake's feet sticking out of this here fish's mouth . . . So I grabbed his lifelines and started to pull hard. Nothing happened. That fish just remained there without so much as blinking one of those big eyes of his. That gave me a sudden inspiration. I hit him square in the eye with one of my fists. He just spit Jake out.'

Mariners' tales are legion, some based on real events and others the fabrication of minds numbed by being at sea for long periods of time. Undoubtedly, strange and sometimes dangerous beasts are lurking beneath the surface of the world's oceans and seas, but the same may be true of smaller bodies of water – in lakes, rivers and even streams. Here there is a whole range of predators just waiting for us to wade in.

A local fisherman in his canoe loses a finger, cattle crossing a river reduced to skeletons in minutes, and somebody bathing in a lake disembowelled while they wash – all horror stories associated with the infamous South American piranha.

That piranhas have a dreadful reputation, there is no doubt, but is it justified? Their name comes from the native 'tupi' language – pira meaning fish, and ranha meaning teeth – reference to the small, sharp triangular teeth that can sever a finger or toe in one bite. In other parts of South America they are known as caribes; the Maka refer to them as wanaj; and in the Guianas they are called perai. Of the 16 species, only a few – including the piranha (*Pygocentrus piraya*), red-bellied piranha (*Serrasalmus nattereri*) and black piranha or pacu (*Serrasalmus niger*) – are considered dangerous to people. Some species specialise in eating nuts and fruit that fall from trees lining the river banks, but it is the flesh-eaters that fascinate people.

The black piranha is the largest with a length of 16in (41cm).

It is also the rarest and the fiercest. Aquarium keepers have a healthy respect for the fish for it will kill and eat any other fish in the same tank, and will even track a person walking close to the glass. In most public aquaria, however, it is the 12in (31cm) long red-bellied piranha that is the commonest. In the wild, all species are found in the slow-moving rivers of tropical South America, such as the Amazon and Orinoco and their tributaries. They swim in large shoals and are ever alert to commotion or blood in the water. They will home in on a target and, darting in to take small bite-sized chunks, strip it bare within minutes. The water literally boils with the frantic activity. There are records of a 100lb (45kg) capybara being stripped of all its flesh in less than a minute and of a large, injured caiman being reduced to a skeleton in no more than five minutes. In the feeding frenzy the piranhas will some-times attack others of their own kind.

When crossing a piranha-infested river in the Pantanal, the local farmers were said to save their valuable cattle from damage by taking an old ox and throwing it into the river. The piranha immediately attacked it, while the farmer moved his herd safely across the river further along the river bank. The story, though, may be no more than a cowboy myth.

Piranhas are active throughout the day, with bouts of intense feeding at dawn and again in the late afternoon. They sometimes stalk prey, other times wait in ambush. They might actively chase a victim or simply circle prey waiting for the right moment to dart in and tear it apart. Whatever they are doing, all the fish in a shoal face the same way, a safety precaution in case a neighbour should turn cannibalistic.

Ferocious piranhas seem to be confined to particular stretches of rivers or lagoons, but not all are man-eaters. In some places people swim in rivers alongside piranhas and emerge untouched. In others, they would be unwise to put even a toe in the water. *National Geographic* scientist Paul Zahl tells of an expedition in the late 1960s to the River of Death, a tributary of the Araguaia, and to a reach with particularly ferocious fish. His guide lowered a dead monkey into the river, the surface boiled, and five minutes later only a skeleton was pulled out.

Water conditions, such as temperature, current speed and salinity in estuaries, might well influence whether or not piranhas attack. Zahl visited Marajo Island in the Amazon delta where a stream packed with piranhas rose and fell with the tide. His host told him that the fish only attack when the tide floods, and went on to prove his point. He took half a goat and tossed it into the water at high tide. It sank to the bottom untouched. A few hours later, he staked out a dead caiman and as the tide turned and the bait was covered by the incoming tide, the piranhas were whipped up into a feeding frenzy. Fish leapt from the water in their haste to get at the carcass. Three minutes later only fragments of hide and the bare bones remained.

Piranhas certainly have the wherewithal to inflict serious injury. They have fiendish razor-sharp teeth, and jaw muscles strong enough to bite off a person's finger like a knife slicing though carrot. When naturalist Nicholas Guppy visited Guyana in the early 1960s, he found that many people at Apoera on the Courantye River had fingers or toes missing, the lost digits having been taken by piranhas when the local folk went to the river to bathe or wash clothes. They also had small chunks taken out of arms and legs. One boy, who lived at nearby Orealla, had most of his foot bitten off by piranhas.

Fatal attacks on people appear to be rare, and are usually the result of other accidents. In September 1981, a passenger-cargo boat capsized beside the dock at Obidos on the Amazon in Brazil. Of the 500 people on board 178 survived, and only four bodies were recovered from the 250ft (76m) deep water. The rest were probably drowned and, according to the local authorities, may have been eaten by piranhas.

It is not unknown for piranhas to scavenge on human corpses. There are several reports of this behaviour from the Matto Grosso in western Brazil. The people there have usually drowned, either as the result of an accident or due to heart attacks. The piranhas take advantage of the sudden arrival of food. In one case, all the soft tissue on the head of a corpse was stripped away completely by piranhas not long after the person had drowned. Another was reduced to a skeleton after four days in the water, and a third body, recovered about 20 hours after a heart attack, had all the flesh removed except for some left on the trunk. The culprits were *Pygocentrus nattereri* and *Serrasalmus spilopleura*. It is thought that the many reports of piranhas devouring people are more probably cases of the opportunistic fish scavenging on drowned or otherwise already dead people.

Piranhas more usually feast on more mundane things. When young they might take small aquatic invertebrates, and when they grow larger change to eating small fish or taking chunks out of larger ones. Some young piranhas specialise in grabbing pieces of fins, hiding below floating water hyacinths and darting out to grab a mouthful. When they are older they mill about under the nesting colonies of birds, such as egrets, waiting for a clumsy fledgling to fall in.

But you do not have to be in the Amazon basin to get a nip from a piranha. Many an amateur fish keeper has had a chunk taken out of a finger while feeding their 'pets', but imagine the surprise of the policeman taking part in a drugs raid in Essex in October 1999 who bent to pick up a piranha that had been spilled on to the floor only to be bitten on the hand for his trouble.

In the Old World, Africa's tiger fish (*Hydrocyanus goliath*) is the largest of the piranha family (the characins). It grows to 5ft (1.5m) long and weighs more than 100lb (45kg). Shaped somewhat like a sleek, streamlined salmon, it differs in having very large, knife-like teeth. The teeth of upper and lower jaws intermesh, giving the jaws the cutting capability of garden-shears. It gets its name from the stripes along its body.

The tiger fish inhabits the open water of main river channels, such as the Zambesi and Congo, and fast-flowing tributaries. Its primary food, when adult, is cichlids, but its ferocious-looking dentition and voracious appetite makes it a potential danger to anything or anybody in the water. It hunts in large packs, much like piranhas, and takes large chunks out of its victims in the same way as its South American relatives. Like many man-eating reports, there are unverified tales of attacks by tiger fish on people.

There are, however, two verified tales of tiger muskies or muskellunges (*Exos masquinongy*) – solitary and relatively uncommon pike of the North American Great Lakes region and elsewhere that grow up to 6ft (1.8m) long – attacking people. In 1995, a 14-year-old boy from Minnesota was swimming on a public beach at the side of Lake Rebecca when he was bitten on the right hand and wrist by a 38in (97cm) muskie, and in 1998 a young man was dangling his feet over the side of his canoe on Twin Valley Lake, Wisconsin, when a 36in (91cm) muskie grabbed one. Jerking his foot out of the water, he brought out the fish too. He decided to keep it and mount it, but only after he paid the $10.55 special permit fee to the Wisconsin Department of Natural Resources because the fish was not caught on rod-and-line and it was under the statutory 40in (102cm) minimum length. His foot needed 60 stitches.

A far cry from the tropical lakes and rivers of South America and Africa or the lakes of North America, is a pond near Ascot racecourse, but it was here that a peculiar man-eating incident took place in 1856. George Longhurst's young son was swimming in the pond, and he was attacked by a large northern pike (*Exos lucius*). In a flurry of spray, the fish grabbed his hands and wrist and realising it would not be able to swallow the rest of the boy, it let go. One of its teeth had passed right through a fingernail. The cause of the attack was evident some days later, for the fish was found dead on the shore. It had starved, and while hunger wracked its body, it was attacking anything that remotely resembled food.

In the rivers of Europe, a solo hunter has been dubbed 'man-eater'. It is the European catfish or wels (*Siluris glanis*). A specimen found in the Dnepr River, which runs into the Black Sea to the east of Odessa, was estimated to be 15ft (4.6m) long and weighed 720lb (272kg). Another in Lake Zwischenahn, in north-west Germany, was estimated to be more than 15ft (4.6m) long. Specimens almost 10ft (3m) long and weighing more than 375lb (170kg) have been taken in Romania's stretch of the Danube, and a 565lb (256kg) one was caught in the Desna River in the Ukraine. Another giant, thought to weigh 661lb (300kg) is claimed for the Dnepr. With a mouth more than 2ft (60cm) across, fish of this size are potential man-eaters, and indeed, reports – though uncommon – exist of people, mainly children, being taken.

There was the story of two young Hungarian girls, for example, who had gone to the river to collect water and were taken by a gigantic catfish. Another tells of a wels caught by a fisherman in Turkey that contained the body of a young woman. Elsewhere, children and parts of adults have been found in the stomachs of catfish, but as catfish feed mainly at night it is thought they were more probably scavenged from rotting corpses that were at the bottom of rivers after drowning or the result of some other accident.

Although the European wels was thought to have been the largest living freshwater catfish in the recent past, the title today should probably be given to the very rare pa beuk, pla buk or plaa buk (*Pangasianodon gigas*) of the Mekong. It lives in the deeper parts of rivers in Laos, Cambodia, Vietnam, Thailand and parts of China. It grows to a length of 9ft 10in (3m) and weighs up to 529lb (240kg), and it is caught for food during April and May by folk living along the river. About 30 or 40 are caught each year, and its flesh is prized as an aphrodisiac. A fish of this size may not be dangerous to an adult human, but like the European catfish, its mouth is large enough to snatch and swallow small children bathing in the river. The chances of it doing so, however, are extremely remote because the toothless maw normally sucks in nothing more than algae.

South American rivers also have their giant catfishes, the longest being the lau-lau (*Brachyplatystoma filamentosum*) which grows to 12ft (3.7m) long. Another is the pirahyba (*Piatinga piraiba*) of the Amazon. President Teddy Roosevelt, when on an expedition in Brazil, was told of a pirahyba near the mouth of the Madeira River which was 9ft 10in (3m) long. It had attacked two men in a canoe, and was dispatched with machetes.

Again, though, it is India that figures in another bizarre man-eating story from Uttar Pradesh. This time the authorities are breeding man-eaters deliberately. In a $140,000 programme, two turtle breeding centres have been established at Varanasi and Lucknow and the task their aquatic reptiles have been set is to help clean up the Ganges. The river is badly polluted. There is untreated domestic effluent pouring in from the sewers or nalas, pesticides and herbicides from the fields, industrial waste from factories, and a whole lot more. It is, however, a holy river – Ganga Ma or Mother Ganges – and Hindu pilgrims arrive from all over the subcontinent just to bathe in the river from the stone steps, or ghats, that line its bank. Many acquire more than spiritual well-being. Hepatitis, cholera and other infectious diseases are rife.

One element of the problem is human corpses. The dead are cremated on the banks of the Ganges and ashes and partly burned bodies are tossed into the river. This ensures moksha – release from the cycle of death and reincarnation. Children and holy men are so pure they do not need to be burned and so their bodies are put in the river intact. Very poor people also launch their dead directly into the river. At one time, sharks and river turtles took care of the bodies, but the pollution has meant they are unable to survive in the filthy waters. The local species of turtle – the 28in (70cm) long *Triunyx gangiticus* – was brought to the edge of extinction. Now, the Varanasi Development Authority wants to put the turtles back in the river in order that they nibble away at the dead bodies.

Eggs are collected from the relatively clean Chamba River, and brought to the two centres. At first the turtles are vegetarian, but at 18 months they turn to meat-eating and that is when they are released into the Ganges to do their gruesome job. More than 600 turtles are tagged and released each year. About 60 per cent remain in the Varanasi region, and many have been observed feasting on the bodies that float by. With the introduction of a more efficient modern crematorium on the river bank and a successful clean-up programme by the turtles, there is the concern now that the turtles might one day run out of corpses and take to nibbling live bathers as they immerse themselves in the sacred Ganges.

And if that is not strange, how about this tale of robber crabs (*Birgus latro*), possibly the most bizarre man-eating story ever told? These large crustaceans climb trees and feed on fruits including coconuts, their large and powerful claws being quite capable of breaking into the tough husks. They are ostensibly vegetarians, so it comes as a surprise to read a newspaper report from 1951 about hordes of robber crabs attacking sleeping Moslem pilgrims who were marooned on an island in the Red Sea. Twenty-six of the pilgrims were killed when the crabs punctured their skulls. Did they mistake the human heads for coconuts? We shall never know.

# CHAPTER 14

---

# MINI-BEASTS AND BLOODSUCKERS

Small creatures make up for their size by sheer weight of numbers. In the insect world, for instance, most species are too small or too weak to attack people, but when many millions of insects get together and work as one – a kind of super-organism – these tiny creatures become potentially dangerous. Ants are the prime movers.

Army and driver ants (*Eciton* and *Anomma*) are potentially man-eaters, although the chances of somebody falling prey to them is very remote. They are basically insect nomads that scour the South and Central American forests and the East African plains respectively for prey. They travel in great columns, like an ancient army on the march, with upwards of half-a-million ants in a raiding party. They catch and slice up just about anything that lies in their path. Fortunately they move slowly, so any healthy individual could out-run (or even out-walk) them. An injured person incapable of escape, or a baby in a cradle, however, could fall victim.

In December 1973, for example, a column of army ants reported to be over a mile (1.6km) long and half-a-mile (800m) wide invaded the town of Goianira in central Brazil, and – if newspaper reports are to be believed – killed and shredded several people who were unable to flee, including the chief of police. The column was apparently driven back into the jungle by 60 soldiers with flame-throwers.

Whether the Goianira event was real, exaggerated or imagined is difficult to establish, but events at an 'ever-so-comfortable' hotel in Tikal, Guatamela, were very real for Mary Mackey and her sister. Mary tells her story in *I Should Have Stayed at Home*. One night, she and her sister were asleep in the hotel. Suddenly, she woke up with a start when a 'host of little things' bit her simultaneously. Her sister went for the light switch but the electricity was off, and so Mary leaped around in the dark trying to brush off the offending 'little things'. Her sister grabbed a flashlight and to their horror discovered that Mary was covered from head to toe with army ants. Using a towel, the two tried to brush and beat the ants away. Flashing the torch around the room they were horrified for a second time to see a column of ants, about 5ft (1.5m) wide and several inches thick, swarming down the walls of the room like water down a waterfall. Eventually, the ants took over the entire hotel, and while the guests – mostly European tourists – huddled together in the reception area, the ravaging columns began to dislodge the hotel's other uninvited guests. Scorpions started to drop from the hotel's thatched roof, and so the human guests sat for the rest of the night with umbrellas up and scorpions falling all around.

Another true ant story comes from Africa. Ivan Hudson and his brother were exploring a mountain on the shores of Lake Malawi. Passing through forests at the base, they walked for three hours taking in the change of vegetation as they climbed. By the time they reach the top, there were few plants and they discovered the summit was a relatively flat plateau. About 328ft (100m) from the top they came to a terrace, which was much easier to walk across than the harsh, rocky terrain over which they had been clambering. Scattered across the terrace were the skulls and bones of small mammals, ranging in size from rabbits to mice, but they did not take too much notice and climbed higher. Then, what had been hard rock underfoot suddenly became a spongy surface. They had walked no more than 20–23ft (6–7m) when they were attacked by red ants. They began to run, but the ants were pouring out of the ground in vast numbers. There were so many, the column looking like a writhing pipe the thickness of a large man's thigh. The mass came out at such a rate it became airborne, and the two walkers were covered with ants. They ran for all they were worth to the top of the mountain, and for no apparent reason the ants simply fell off, scuttling back to the spongy ground below. There seemed to be some invisible line beyond which the ants would not pass.

At that moment, Hudson realised why there were so many skulls on the terrace below the spongy ground. When the ants attacked, the quickest way out was over the edge and on to the rocky terrace. An animal covered in ants and in a state of abject panic would be not too careful where it put its feet and most probably fall. Lying dead or injured on the terrace floor, the ants would be able to feed at leisure on the newly won prey. The ants, Hudson surmised, were the infamous 'Seruwi' ants, a meat-eating species known for the way they invade chicken runs and strip the inmates of their flesh until they are just a pile of feathers. An injured climber on Hudson's mountain would likely suffer the same fate.

If all this sounds far-fetched, consider a paper in *Annals of Internal Medicine* in which Ricard deShazo of the University of Mississippi reports that swarms of fire ants have been attacking bedridden residents in nursing homes throughout the south-eastern USA. The ants arrived in the USA from South America in the 1930s, and they are now widespread. They climb the legs of the bed and cover the victims. Some suffer allergic reactions to the ants' stings, and one person has even died. The ants, according to deShazo, have identified the slumbering people as a potential source of food.

Other insects can be considered part-time man-eaters. Rather than devour us whole or run off with a limb, they eat a small part of us – maybe our tissues or our blood. The tissue-eaters include small flies, although it is the larvae rather than the adult that has taken to man-eating. The grubs of the human bot fly or torsalo fly (*Dermatobia hominis*) in the tropical

Americas and that of the tumbu fly (*Cordylobia anthropophaga*) in Africa, for example, can live on people.

The life cycle of the bot fly is unusual. It is a noisy flyer and would instantly draw attention to itself if it tried to approach a victim directly. Instead, it seeks out a quieter flyer. It captures a blood-sucking insect, such as a blood-sucking fly, tick or mosquito and glues its eggs to the underside of the body. When the blood-sucker goes to feed on its warm-blooded host, the warmth of the body causes the larvae to hatch out very rapidly, and they burrow into the host's skin. They feed on the living tissue underneath, producing swollen, boil-like lesions or myiasis which can be very painful. Some tunnel into the nostrils. Here they stay for about six weeks, after which they emerge (or are sneezed out in the case of nasal invaders) and pupate in the ground.

There is a wonderful story of a distinguished biologist playing host to the grub of a bot fly. The larva was below the skin at the back of the man's neck. Rather than remove it, he decided to wait for the adult fly to emerge so that he could identify it, but it chose to do so just as he was checking in at the airport. The women behind him fainted, and in the confusion the fly broke out and flew off. He never did discover which species he had been carrying!

The lesions produced by bot fly larvae may also attract other myiasis-producing flies, such as screwworms (*Cochliomyia*). The female screwworm – which is actually a fly – deposits her batches of 10–400 eggs at the edge of a wound or even a tick bite. The larvae hatch and feed on any blood or secretions from the wound and on the living flesh. Tissues are liquefied and the lesion can extend significantly. If the condition is not treated, the victim can die.

A particularly nasty African fly is *Callitroga hominivorax* (which means quite literally 'man-eater'). Its larvae enter the nose, chew their way through the nasal bone within hours, and into the brain. Death is usually from meningitis.

Some invertebrate hitch-hikers do not kill their hosts but they can be a nuisance. Sand fleas or jiggers bury themselves in the host's skin, usually between the toes. Mites can be irritating too. Larval mites known as chiggers feed on lymph, and their saliva can cause an allergic reaction on the skin of some people. Scabies or itch mite also infect humans.

Some miniature 'man-eaters' are actually useful. The follicle mite (*Demodex follicorum*) is part of the life that lives permanently on people. It exists in the hair follicles and sebaceous glands of the skin, where it helps to keep the skin clean by consuming human sebum. The house mite consumes dead human skin cells, but its usefulness is tempered by the way its own moulted skin particles induce allergic reactions, such as asthma, in people.

The most dangerous invertebrates, however, tend to be the blood-sucking insects. Mosquitoes, midges, gnats, black-flies, horseflies, fleas and bed-bugs, are all consumers of human blood, and many also carry dangerous and debilitating diseases.

The mosquito (*Anopheles*) is probably the most dangerous insect in the world for it spreads malaria, yellow fever and a whole host of other diseases in humans and is responsible for the deaths of many millions of people every year. It is not the mosquito which kills, of course, but the parasitic diseases it carries – malaria is caused by the protozoan *Plasmodium*, for example. The insect itself is only interested in a meal of blood which it obtains by piercing the victim's skin with mouthparts resembling a hypodermic syringe.

The mosquito lands on its host so lightly that it cannot be felt, and then feels around the skin with a pair of palps to find a soft spot. The piercing mouth stylets are then pushed down, with a hypopharynx at the centre. In a minute longitudinal groove in its dorsal surface saliva flows down, while the blood flows up another groove on the underside of the upper lip. Only the female mosquito has this arrangement of mouthparts and gorges on blood, the boost of protein utilised to make more eggs.

The tiny midge (*Culicoides*) or 'punkie', as its known in the USA, has similar mouthparts to the mosquito and can be a troublesome nuisance. Clouds of these tiny creatures can attack any exposed part, making it very uncomfortable for people in some parts of the world to sit out on a summer evening. In the western highlands and islands of Scotland, for example, the midge *Culicoides impunctatus* can make life a misery for tourists and locals alike. One British zoologist reckoned that 'its presence in conjunction with that of the kilt is said to have given rise to the Highland Fling'.

The horseflies and cleggs (Tabanidae) obtain their blood meal, not with elongated needle-like mouthparts like the mosquitoes and midges, but with shorter, blade-like mouthparts that can inflict a painful bite. They can be extremely persistent, taking bites while avoiding being swatted.

Fleas are specialised blood-suckers. They evolved from normal flying insects, probably the scorpion flies, about 160 million years ago when the mammals were still in the shadow of the dinosaurs. It is probable that fleas and mammals evolved together, and they tend to focus on a particular host, although not exclusively. The cat flea, for example, may feed on the pet's owner, but it would prefer to be on a cat.

Whatever the flea and whatever the host, the flea's method of extracting blood is remarkable. Inside its head is a membrane made of an elastic protein called resilin. Next to this is a hammer-like bar attached to the insect's piercing mouthparts – the stylets. When the flea reaches a suitable patch of skin, the stylet muscles push the bar against resilin membrane, much like an archer pulling back the bow-string when preparing to fire an arrow. The flea tilts its head down and its body up, and relaxes the stylet muscles, like the archer letting his bow-string loose. The resilin membrane springs back, propelling the bar and the stylets into the host's skin. This is repeated rapidly (now more like a jack-hammer than an arrow) until the flea reaches a blood capillary and begins to feed.

Fleas and other blood-sucking insects, such as bedbugs, ensure they extract a bedbug-sized meal with the help of a protein in their saliva. The protein, known as nitrophorin 1, not only dilates the victim's blood vessels but also mops up the histamines produced by the surrounding cells (they cause the itching sensation). It works when the insect pierces the skin with its needle-sharp proboscis and injects saliva, containing the protein, into the tiny blood vessels there. Ann Walker and her colleagues at the University of Arizona in Tucson discovered that the protein stores nitric oxide, which is known to dilate blood vessels. As it enters the host, there is a drop in acidity and this triggers the release of nitric oxide. Thus, the protein enables the bedbug to enjoy a big meal without alerting the host.

Fleas, like other blood-sucking insects carry dangerous diseases, such as plague, and lice, another group of insect parasites. They spread typhus and relapsing fever. The insect responsible is the human body louse (*Pediculus humanus*). It is one of three lice that are frequently found on people, the others being the crab louse (*Pthirus pubis*) which targets human pubic hair, and the ubiquitous human head louse (*Pediculus capitis*) whose infestations are heavy amongst school children.

Lice grip the hair of their host with modified claws at the end of each leg, and pierce the skin with three stylets. They then take a blood meal. In Sweden, lice had an influential position in choosing councillors. The town or village elders would sit around a table and place their substantial beards on the table. A louse was placed in the middle, and the owner of the beard into which it made its home was appointed the next mayor.

Other invertebrates that have taken to blood-sucking are the ticks and mites (Acari), parasitic relatives of the spiders. Ticks are the most conspicuous, being larger than mites, and they are frequently found on people. In order to receive an early warning of an approaching host, the tick possesses a sophisticated sensory system. In special depressions in its cuticle (skin), it has concentrations of hairs sensitive to humidity and odours, and on the end segment of both front legs are special detectors. When a host is detected the tick heads to the top of the nearest grass stalk or other convenient vegetation and waves its front legs in the air in order to manoeuvre its body into the right position to make a grab. It waits for its victim to brush past.

Successfully aboard, the tick punctures its host's skin with the serrated pincers in its mouthparts, and pushes a snout-like projection, called the hypostome, into the wound. Teeth on the hypostome help anchor it in place, and a groove along its upper surface transfers the blood into the tick's mouth. Anti-coagulants in the tick's saliva help stop the blood from clotting. The tick may stay in place for several days, growing fatter and fatter. Blood is a bulky food, but because meals are few and far between, ticks gorge themselves in one sitting. A female tick might store 200 times her own body mass (from 2mg to 400mg) in her digestive tract during the seven or eight days that the food lasts. The tick then drops off to moult or deposit eggs.

Ticks can also be dangerous because, like other blood-suckers, they carry dangerous diseases. The deer-tick, for example, carries the bacteria that causes Lyme's disease in humans. Others carry viruses like those responsible for spotted fevers, tick-typhus fevers, haemorrhagic fevers or encephalitis.

The tragic disease rabies is spread by another infamous but somewhat larger blood-sucker – the vampire bat. Vampire bats are exclusive consumers of blood. They were given their name – 'vampir', the Magyar word meaning person who comes back from the dead and drinks the blood of living things – by the Europeans who invaded South and Central America in the 16th century. There are three known species, and they are not huge flying creatures as depicted in horror movies. Each weighs a maximum of 1.4oz (40g) and has a wingspan of no more than 20in (50cm).

People are not usually the main target. While the hairy-legged vampire (*Diphylla ecaudata*) prefers to prey on birds, domestic stock is preferred by the other two species, the common vampire bat (*Desmodus rotundus*) and white-winged vampire bat (*Diaemus youngi*). Humans are a fall-back alternative when cattle, donkeys or horses are not readily available. People asleep, who have left a foot dangling outside their mosquito net or a cheek, forearm or nose exposed, are sometimes attacked by vampire bats.

The bat approaches by stealth, using echo-location to locate likely targets, so as not to alert its victim. It detects warm-blooded prey with the aid of heat-seeking pits in the nose. Thermo-receptive areas in the nose-leaf can detect infra-red radiation. It has two razor-sharp incisor teeth with which it takes a shallow slice out of the skin. Contrary to popular belief, it does not 'suck' the blood, but licks it up with a specially grooved tongue. Anti-coagulants produced in the bat's salivary glands prevent the blood from clotting. It takes about 10cc from each victim, and urinates constantly while feeding. This enables the bat to rid itself of the non-nutritive parts of its meal and so lighten its load when the time comes to take off. The tubular (intestine-like) stomach enlarges considerably to accommodate the meal. The attack is so sneaky, often as not, that the victim does not wake up and is unaware that the bat has been taking his or her blood.

Interestingly, horses and cattle can detect the calls of vampire bats and move to places, such as the open prairie, where the bats are less likely to attack. They also keep moving. The bats prefer to be near vegetation where they can hide, and tend to home-in on animals which are still or are lying down. The bats themselves fall prey to domestic cats.

Some of the smaller blood-suckers attack from the inside rather than the outside. The hookworm larva, for example, enters the human body through the skin of the feet, and makes its way to the gut. Here, it attaches itself to the wall of the stomach with hook-like teeth. It bores a hole and sucks the blood. Chemicals are injected to prevent the blood from clotting, and so when the hookworm moves on, the wound continues to bleed. People with particularly bad infestations of hookworm are often anaemic.

Another commonly encountered blood-sucker is the leech, an external parasite of mammals, fish, amphibians and birds. There is not just one leech but over 300 known species, some of which have an affinity for human blood. The leech waits for a passing blood meal in damp vegetation. It has a large sucker at the rear end with which it can anchor itself to a leaf, and a smaller sucker at the front end which it can attach to the victim. Once attached to the host, the leech uses its three jaws containing sharp teeth to make a Y-shaped incision in the skin. Its saliva anaesthetises the area of the wound so that the host barely feels a thing and it causes the blood vessels of the host to dilate so that the blood flow is greater. It also contains an anti-coagulant to stop the blood from clotting. Often a person only knows he has a leech on board when he sees the trickle of blood running down his leg.

Physicians in Europe used the blood-sucking ability of the medicinal leech (*Hirudo medicinalis*) to drain off blood. It was popular as a treatment for all manner of diseases during the 19th century and then fell out of favour. Today, the leech is being used again by surgeons. It is an effective and precise way to drain a blood-clot or control internal bleeding.

Some leeches, such as small or young aquatic species, actually enter the bodies of people who bathe in infested waters. They gain entry through excretory openings and suck the blood from the inside. If they are in drinking water, they can infest the nasal passages, lungs and linings of the throat. If there are too many, a person can suffer from anaemia from loss of blood and die.

Another underwater but unexpected blood-sucker is a small species of fish, known as the candiru (*Vandellia cirrhosa*). Its uncomfortable habit was discovered in 1994 by Wilson Costa, of the Federal University of Rio de Janeiro, in the Araguaia River in the south-east of the Amazon Basin. The fish is transparent and just 0.3in (1cm) long. It normally enters a host's body, such as the gills of a fish, where it feeds exclusively on blood. It has two hook-shaped teeth at the back of the mouth and a flexible jawbone with which it latches on to its prey and draws blood. Unfortunately, the fish can also slip into human orifices where it is almost impossible to shift. On one occasion, one of the tiny fish slithered into a cut on a researcher's hand and could be seen wriggling under the skin, making for a vein. It has gained notoriety in some parts of Brazil because, when a male bather is in the river, it has the unpleasant habit of swimming up the urethra of the penis and is almost impossible to dislodge. Spines on the gill covers anchor it in place. The fish can reach the bladder where it feeds from blood vessels in the wall. Secondary infections can occur and the unfortunate person may die.

# CHAPTER 15

---

# CANNIBALS, SACRIFICES AND MAN-EATING TREES?

## CANNIBALISM

People eat people. The prevalence of cannibalism in humans has been a controversial issue amongst biologists, anthropologists and archaeologists, but the inescapable truth is that, as a species – like sharks, frogs and a host of other animals – we are capable of eating our own kind.

Skulls of primitive Peking Man, dated half-a-million years before the present, were found with the bases enlarged to allow better access to the brain. Limb bones were smashed and splintered, most likely broken for the marrow inside. The perpetrators were not thought to have been wild animals but other people. Whether cannibalism was part of some early religious ritual or whether advantage was being taken of some fresh meat after a fatal accident or something similar is not clear.

Evidence from Krepina, in the former Yugoslavia, suggests

cannibalistic feasts were taking place there over 100,000 years ago. Some authorities suggest this was a way in which to pay homage to the dead, rather than ritually humiliate an enemy. The Issedones, who lived in Central Asia in 500BC, ate their fathers, liberally mixed with lamb, and it is claimed the ancient Irish feasted on their dead. To this very day, the Amahuaca in Peru grind up the teeth and bones of their dead and make up a potion which is consumed by the relatives.

In the Middle East, there were old men who gave their bodies in order to save others. They stopped eating, except for a diet of pure honey, and when they died they were put in stone coffins filled with honey in which they marinated for a hundred years. The confection was used for the treatment of wounds. In Persia, the victim was not so old. A 30-year-old was drowned in a mixture of honey, drugs and herbs and then sealed up for 150 years. When he reached maturity, his body was used as a sweet medicine.

And, as late as the early 20th century, the preserved dead flesh from Egyptian mummies was used as a cure-all – for epilepsy, rashes, stomach pains, ulcers, liver disorders, bleeding, bruises and poisoning – throughout Europe. At first, it was pure Egyptian flesh that was used, but as apothecaries realised the market value of mummy flesh, all sorts of dead humans and animals were dried and ground up, leading to the spread of all manner of diseases. Such was the demand for human remains, cups of blood and lumps of fat were on sale immediately after an execution.

Many recipes from the 16th and 17th centuries made full use of a fresh and clean corpse. Oswald Croll extolled the virtues of 24-year-old unblemished males in his *Basilica Chymica*. The body was to be left in the open air for a day, the flesh cut into strips and seasoned with myrrh and aloes. It was marinated in wine for several days, then dried. It was advised to marinate it again in olive oil before eating to avoid stomach ache. It kept the plague away, amongst other things, or so it was claimed.

Another recipe created 'a most efficacious balsam'. This was Nicolas Le Fevre's 'mumiall balsam'. Again, the muscles of a 'lusty, young man' were required. They were cut up into strips, marinated in wine and then air-dried and smoked with burning juniper wood. The dried strips were put in a jar with olive oil and turpentine, and steamed in a hot bath for 40 days. Further manipulation with theriac and powdered viper's flesh eventually produced a balsam suitable for the treatment of poisons and palsy.

Human brains were also popular. In a London pharmacopoeia dated 1682 enthusiasts were encouraged to take the brain of any young man who had suffered a sudden and violent death and grind it up with a concoction of flowers and fruits and distil. It was supposedly good for epilepsy.

In some parts of the world, such as the New Guinea highlands, the ritual consumption of dead relatives was not only a mark of respect, but also provided vital protein in a high-bulk but protein-deficient diet. The highlanders ate the brains of recently deceased tribesmen.

The belief that humankind can be cannibalistic goes in cycles. At the beginning of the 20th century, the scientific community believed that people were basically rather nasty and that some human societies condoned the ritualised slaughter and consumption of their fellow human beings. By the 1960s, the fashion had changed. Archaeologists considered that the scratches on bones were the effects of weathering or the teeth marks of scavenging animals. But then some scientists began to question this view.

In 1967, anthropologist Christy Turner, of Arizona State University in Tempe, discovered that 30 members of the Polacca Walsh tribe in Arizona had been butchered and eaten. Their bones showed signs of having been cut, burned and broken open, and this clearly indicated to Turner that these remains were 'the leftovers from a cannibalistic feast'. He came to this conclusion by taking a variety of animal bones and subjecting them to the ritual processing – cutting and burning – that the Polacca Walsh bones would have received. Comparing the old bones with his experimental ones, he saw that the marks were identical. The people responsible for eating their neighbours were the Anasazi, the ancestors of the Hopi and Pueblo Indians. They did not eat them because they were hungry, but used the threat to terrorise other tribes. The Aztecs did the same in South America during the 16th century. Several Aztec temple sites contain the splintered human bones that suggest ritualised cannibalism. And, at the excavation of a Spanish settlement dated 800,000 years ago, evidence of human bones stripped of flesh indicate cannibalism was present there too. The Aztecs prepared their human flesh, according to Spanish invaders, in a tomato and pepper sauce.

Samuel Hearne, journeying in North America during the mid-1700s, wrote about Cree Indians eating human flesh. It was said by the Cree that if any of their number turned to cannibalism, then no person was safe in their company. The cannibal would not be trusted and nobody would tent near him. Hearne describes how an Indian came to him when he was building Cumberland House in the spring of 1775. Other Indians, living at the plantation, questioned him and the women checked meat he was carrying in his bag. The general consensus was that he had killed a person and was carrying his flesh to sustain him on his journey.

Jared Diamond, of the University of California, Los Angeles, writes about a more recent example of cannibalism in his book *Guns, Germs and Steel*. In 1835, when the Maoris invaded the Chatham Islands (Rekohu) to the east of New Zealand, they not only enslaved the local people, but also killed and ate them. One Maori was quoted as saying, 'it was in accordance with our custom'. It was the ritual humiliation and finishing off of the enemy.

Access to the power of the enemy through cannibalism was common in many parts of the world throughout history. Pre-dynastic Egyptian soldiers indulged, as did the Chin armies of China, the Scythians in the Mediterranean region, the Uscochi in the Balkans, the Iroquois of north-east America and the Ashanti of Africa. It was said extremists in Nazi Germany ate human flesh during the Second World War, and more recently, in 1971, a Black September terrorist claimed to have drunk the blood of the Prime Minister of Jordan whom he had just assassinated.

Cannibalism also featured in political ceremonies and religious rituals, as a symbol of power and a means of suppressing lesser mortals. Shih Hu, a Chinese ruler who came to power in 3334BC, enlivened his feasts by selecting a young woman from his harem and beheading her. Her body was served for dinner, but her head was left uncooked so that his guests could see that he had not selected the least attractive girl.

On the eastern side of the Pacific, the Easter Islanders were 'kai-tangata' or man-eaters. Women and children (who were sometimes the victims) were excluded from the ceremonies, but the men of the islands headed for secluded places and indulged their passion for human flesh. Captives were imprisoned and then let out at the moment of sacrifice. The practice was not only religious. The islanders actually liked eating people; after all, humans were the only large mammals on the islands with readily available fresh meat. Apparently, fingers and toes were the greatest delicacy.

Indeed, cannibalism was rife on islands right across the Pacific, much to the consternation of Christian missionaries in the 19th century. The Reverend David Cargill wrote of a visit he made to Rewa in Fiji during which cannibalism was considered an everyday event. On one occasion a visiting tribe brought the bodies of 20 men, women and children as goodwill gifts. Some were prepared as food, the entrails and heads floating past the mission station. Some bodies were mutilated, the children of the host tribe playing with the body of a dead girl from the visiting tribe. Five years later a Reverend Jaggar visited the island and despite the intro-duction of Christianity, things had not improved. He reported that one of the servants of the king had run away but had been caught. As a punishment, her arm was chopped off below the elbow and the king had it cooked. He ate it as the servant watched. On another occasion, according to Jaggar, the Bau chief ordered two men, sentenced to death for an unknown reason, to dig their own oven and cut the wood for it. Some of their blood was drained and consumed in front of them. Their arms and legs were cut off, but the men were kept alive, so that they could be fed on their own flesh before having their tongues drawn out with a hook and sliced off. The tongues were eaten, and then, before the men passed out, their stomachs were slit and their bowels spread out before them.

Death followed soon after.

On our own doorstep, cannibalism has featured in early British history. In the Cheddar Gorge in Somerset, Chris Stringer of the British Museum (Natural History) has also found signs of cannibalism amongst 12,000-year-old remains. People had been scalped, beheaded, had their tongues cut out and their bones smashed open to extract the marrow. Was this a ritual, wondered Stringer, or was it simply hunger?

People in desperate situations have resorted to cannibalism just to avoid starvation. Such an occasion was in the late 1800s, during the last days of the French Flatters mission at Hassi Hadjadj in the Sahara Desert. The mission camels had been stolen and there was no means of travelling to obtain water or food. Members of the mission killed and ate each other.

A report to the British Admiralty on the fate of the Franklin Expedition of 1845 by Dr John Rae, which was published in 1854, stated that 'from the mutilated state of many of the corpses and the contents of the kettles, it is evident that our wretched countrymen had been driven to the last resource – cannibalism – as a means of prolonged existence'. There had been rumours of cannibalism on Franklin's overland expedition of 1819 to 1822, and the six survivors of the Greely expedition of 1881 to 1884 returned with stories of cannibalism.

In 1846 an expedition led by George Donner, which planned to cross the Utah Mountains to California, came to grief at the highest point in the journey. After crossing a salt desert, a series of murders and an accidental shooting, the remaining expedition members were caught in storms in the mountains. Their cattle and horses were left roaming and the food rapidly ran out. People started to die and instead of burying them, survivors ate their fill and packaged the rest of the human flesh to eat on the journey. Of the 84 people who started out, only one man, a boy and five women survived by eating other members of the expedition.

In 1884, a report in *The Times* told of three shipwreck survivors having eaten their cabin boy, but the most recent example was in 1972, when the Uruguay rugby team crashed in the Andes. The survivors fed on jam, chocolate and wine for the first few days, but with no rescue in sight, some turned to eating the bodies of the dead. Those that did, lived and were eventually rescued. The dramatic story was turned into a motion picture with the title *Alive*.

Less well known is the story of Martin Hartwell who crashed his plane in the Arctic in the same year as the Andes tragedy. He was taking two patients accompanied by a nurse to hospital. The nurse died in the crash and one of the patients soon after. The other patient lived for a short time, and he and Hartwell ate corned beef, sugar cubes and then candles and soap from the nurse's bag. Eventually, Hartwell was the only survivor and he ate portions of the nurse to keep going. He was saved a week later.

During the days of Stalin's Russia, famines in some regions were staved off by the judicious consumption of human flesh, and it continues to this day. In 1996, a man went into a police station in the town of Berezniki, in Perm Oblast, and placed a package of human flesh on the counter. He had bought it from two men on the street. Criminals were trading in human flesh. An investigation by the local constabulary established who the two sellers had been. Apparently, they killed a third colleague during a drinking binge and dismembered the corpse. The mother of one of the miscreants cooked up the best bits for their supper, and the rest was packaged up and sold. The victim's head, hands and feet were later found in the attic.

Similarly, in Chita, four packages of meat which had been heat-sealed in polyethylene were found in a rubbish bin. The finder sold one of the packages and it turned out to be human flesh. Clearly, in places where meat is scarce, human flesh has been seen as a viable alternative. In 1997, the half-eaten body of a man was found in a flat, and his 73-year-old wife was arrested for cannibalism. Three months later two soldiers ate a comrade, and in the prison at Semipalatinsk four cell-mates plotted to kill the next new convict assigned to their cell and eat him. This they duly did, taking slices from the body and searing them on a hot plate or boiling them in an electric kettle. In Moscow, a mentally ill person killed a lodger in his mother's apartment. He took flesh from the victim's forearm and fried it up and ate it.

The first American accused and convicted of cannibalism

was Alfred Parker. In 1873, he was prospecting for gold with several colleagues in the San Juan Mountains of Colorado, when they were trapped in a shack during a series of blizzards. He survived by eating their dead bodies.

The ritual eating of people is not confined to neighbouring tribes. Murderers sometimes eat their victims. Wives and girlfriends, for example, might get back at a violent spouse by killing him and eating the body. Such is the case with a 41-year-old woman from Alamosa in the Rocky Mountains, who was charged with shooting her 51-year-old boyfriend and then cooking up his flesh. The man's torso was found in a closet and his legs, with the meat and flesh carefully removed, were found in a garbage can nearby. Bite-size chunks of human flesh were discovered in a cooking pot.

Then, there are out-and-out serial killers. In 14th-century Scotland, Sawney Bean dropped out of society, moved to the mountains and lived with his wife in a cave. Over a period of 25 years, Bean aided by his wife, children and many grand-children, hijacked passing travellers and murdered them for food. His wife picked through the best bits, preserving them in brine. Eventually they were caught, the men dismembered and left to bleed to death, and the women and children burned at the stake.

In more recent times, there is the case of Arthur Shawcross who ate two Vietnamese girls while fighting in the Vietnam War. He was imprisoned, but released in 1988. He spent the next 20 months killing and eating another 11 women.

In the USA, Otis Toole and Henry Lee Lucas went on a killing spree, and murdered 500 people. Toole cooked parts of their victims for dinner. In Russia, human flesh gourmet Nikolai Dzhurmongaliev killed over 100 women and cooked them up in traditional Kyargyzstan-syle at dinner parties for unsuspecting friends. And in Wisconsin, Ed Gein – the real-life inspiration for Buffalo Bill in Thomas Harris's *Silence of the Lambs* – killed and ate many people, including the 15 victims police found in his house. He also turned their body parts into clothes and soup bowls.

In places where cannibalism is or was rife, human meat is called 'long pig' or 'hairless goat'. Choice cuts varied from different parts of the world. The Nigerians liked fingers and palms, Easter Islanders opted for fingers and toes, while Fijians preferred hearts and thighs. Preferred methods of cooking a person are also different. In Africa, pots of water are favoured, while in New Guinea they like to steam-cook. Clay ovens are more common in the South Pacific, and charcoal pits are most common in the Americas. Although a critic of the current view on human cannibalism, William Arens, of State University of New York at Stony Brook, asks, 'Where are the recipes?' He feels that the evidence for widespread ritual cannibalism is slim, and that 'cannibalism exists more in the limited culinary imagination of the observer than in the native appetite'.

Nevertheless, we are not alone in the animal kingdom. Many animals are cannibalistic, including our nearest relatives, the chimpanzees. Jane Goodall's study groups in the Gombe National Park in Tanzania have indulged in the odd bout of cannibalism. Males of one group will kill and eat the older males of another. And one chimpanzee mother and her daughter killed and ate several infants that they had snatched from their mothers. They killed their victim by biting through the skull, and then 'the carcass was consumed . . . slowly and with relish, each mouthful of meat chewed up with a few green leaves'. One explanation for this seemingly strange behaviour is that it is motivated by feeding competition. By removing the offspring of others, there is simply more food for the survivors.

Chimpanzees are indeed opportunistic carnivores, taking mainly other primates such as colobus monkeys, young baboons and anything else they can purloin. They do not stop there, however. They have also been known to take and eat human babies. In her classic book *In the Shadow of Man*, Goodall refers to two known instances of chimpanzees kidnapping African babies. In one case, a male chimpanzee was discovered with a human baby whose limbs had been partially eaten.

In reality, cannibalism is relatively rare in the animal kingdom, and appears to occur only under exceptional circumstances. David Pfennig of the University of North

Carolina at Chapel Hill looked at the behaviour and concluded that there is the risk of getting diseases and parasites that co-evolve with their host when eating others of your own kind. This was certainly true of the New Guinea highlanders – the Stone Age-like Fore tribe – who passed on the brain disease known as kuru as a consequence of a fondness for eating the brains of their dead relatives. Kuru was known by the Fore as 'the laughing disease', and it became apparent in the 1950s. At first a sufferer would begin to walk unsteadily, then start to shake uncontrollably. Speech would deteriorate into a slur and emotions were erratic. The face would contort and the mouth form a horrible grin. Eventually, the victim died.

More women and children succumbed to the disease than men and it was confined to the Fore and not evident in neighbouring tribes. A clue as to the cause came when small holes were found in the skulls of kuru victims. Medical researchers Carleton Gajdusek and Vincent Zigas discovered that in the ritual cannibalistic feasts of the dead, the men ate the muscle tissue of their dead relatives but the women and children were given the brains. They found that the disease was passed from person to person when the brain tissue was consumed. Later it was established that abnormal proteins or prions were responsible and that the disease was spread not by a living organism, such as a bacteria or virus, but by a non-living protein. The sheep disease 'scrapie' and 'mad cow disease' are transmitted in the same way. Gajdusek was awarded the Nobel Prize for his work.

## SACRIFICES

One way in which people may end up on a man-eater's dinner table is when they are deliberately fed to the animals. A common method of dealing with the powers of nature is to worship them; and with worship comes human sacrifice.

In many cultures, particularly those in which the people are living close to nature, top predators are often revered as spirits or gods, some seen as benevolent and others as wicked.

Curiously, while the wolf in the northern hemisphere is considered the embodiment of evil and is hated, the tiger and crocodile in Indonesia are a power for good and are treated with respect.

The shark, wherever it is worshipped, has been considered a vindictive god or an artful devil. In certain primitive religions, such as those of the Pacific islands, men became sharks and sharks became men, and the legends have been handed down from one generation to the next in song and dance.

Even today, Hawaiian dance-masters or kanakas keep the legends alive, performing their dances at the sites of ancient shark temples. One tale tells of Kamo-hoa-lii, the king of all the sharks. One day, he met a human maiden and fell in love with her. Her name was Kalei, and Kamo-hoa-lii was so smitten with her that he changed himself into a man. They had a son, Nanaue. He looked just like any other native child, except he had the image of a shark's mouth on his back. Kamo-hoa-lii changed back into a shark and returned to the sea, but before he left, he said that Nanaue must not eat the flesh of an animal or his shark ancestry would come to the fore. Unfortunately, Nanaue was fed some meat and he then discovered how to change back and forth between man and shark. As a shark, he killed many islanders, but he was caught and his body taken to a hill in Kain-alu. The hill was called Puumano, meaning 'shark hill'. The people cut down bamboo from the hill and fashioned sharp cutting tools from the stems. With these they cut up Nanaue. But the shark gods were angry, and to this day the bamboos of Kain-alu are weak and cannot any longer be cut into knives.

Shark worship was more than songs and legends, however. It also involved more dubious practices. Sharks were not to be satisfied by the occasional man, woman or child that they snatched from the shallows. Shark gods required human sacrifices. Priests would go among the crowd followed by an acolyte carrying a noose or similar device. The noose would be cast and an unfortunate bystander ensnared. The noose was pulled tight and the victim strangled. The body was cut up and thrown into the sea in order to appease the shark gods.

Hawaiian kings threw living people into specially built enclosures containing sharks, and gladiatorial contests were staged between people and sharks that had been starved. The enclosure was a semicircle of lava stones enclosing an area of about 4 acres (1.6 hectares) at the edge of the sea. There was an opening to the sea where sharks could be lured in. During a contest the entrance was closed off. The gladiator was equipped with nothing more than a shark-tooth knife – a stick with a shark's tooth at the end. When the shark rushed in for the attack, the gladiator had to swim quickly below and try to slice open the shark's stomach with the single tooth. He only had one chance for, if he missed, the shark would return with its mouthful of razor-sharp teeth.

One such shark temple was destroyed when the US Navy base at Pearl Harbor was built. Curiously, a dry dock constructed on the actual site of the shark pen was destroyed during an underwater volcanic eruption. Local people were not surprised. It was the Queen Shark. She was angry, they said, and she was flexing her back.

Sacred shark pens were built also in the Solomon Islands. Sacrifice victims were placed on altars made of stone, and during the course of mystical ceremonies their bodies were sliced up and thrown to the sharks. Some – the local species – were good sharks, incarnations of the tribe's ancestors, while others were evil. The latter could be driven away with the help of a small wooden statue representing the good, native sharks.

The Hawaiian islanders wore a good-luck charm until quite recent times. Women displayed a shark tattoo on their ankles. This was in remembrance of an ancient Hawaiian princess who had been bitten on the leg by a shark but had survived. The tattoos were to protect women against shark attacks.

The practice of throwing people to the sharks has not been confined to civilisations that live beside the ocean. At Lake Nicaragua in Central America, bull sharks that travelled up rivers from the sea and entered the lake were offered human sacrifices. One enterprising Dutchman, according to an account in François Poli's *Sharks Are Caught at Night*, took advantage of the ritual. The sacrificial bodies were adorned with sacred jewellery, and so he caught the sharks that had been consuming them and amassed himself a fortune. Unfortunately the local people found out and killed him before he could spend it. He was not worthy to be thrown to the sharks. They simply slit his throat and burned his house.

There is no doubt that large and powerful predators, such as sharks, command respect, and were often used as repressive symbols by religious leaders to help control the masses. They could be a real threat too. Sharks were encouraged to congregate close to prisons in order to discourage escapees. In the 19th century, for example, the British authorities maintained penal colonies in Tasmania. Guard dogs and armed guards patrolled the landward side of a prison sited on a narrow peninsula, but prisoners were swimming to freedom. The prison governor ordered all the rubbish to be dumped in the sea around the colony, and the sharks quickly learned about the free hand-out each day. Frantic screams in the night revealed that the newly recruited 'guards' were proving their worth. The number of escapes rapidly declined. Similar stories were linked to Devil's Island and Ile Royale, off French Guyana, and Alcatraz in San Francisco Bay.

## CIRCUSES AND ZOOS

In some cultures, ritualised killing has been taken even further. Throughout history, there are records of societies actually breeding and training wild animals as man-eaters. In Roman times, the Christians were supposedly put in the arena with lions, leopards and tigers and then attacked and devoured. More likely the beasts, having been shut away in dark cages, were terrified on entering the arena and just wanted to return to their cells. It seems that the only way that sport could be had was when animals selected for their ferociousness were actually trained to attack and eat people.

Oriental courts followed a similar tradition. The King of Pegu, whose realm included lower Burma, kept lions, tigers and leopards and, for sport, let them loose with convicted criminals. The Incas kept pumas in dungeons for the same purpose.

In Rome, during the 13th century, a lion pit was established near the Capitol in which troublesome citizens were thrown. Ludwig IV of Bavaria conquered the city and threw a monk to the lions. The pit was closed down in 1414, when a lion escaped and killed a child.

This desire of people to be frightened by potential man-eaters – but at a distance – was to repeat itself many times throughout history. In 1459, for example, the people of Florence revived the Roman circus. They barricaded the entrances to the Piazza della Signoria and released 20 lions, a pack of wolves and some bulls. The well-fed animals eyed each other carefully, but after walking round the arena for some time, they all went to a shady spot and lay down together. The exercise was repeated in 1514. Four cardinals in disguise came up from Rome to watch the spectacle. This time bears, stags, bulls, leopards, horses, dogs and two lions were let into the arena. One of the lions took a swipe at a couple of dogs and killed them, but left the other animals alone. Men hidden in a model porcupine and tortoise went around the square prodding the animals with spears. As a consequence, several people were injured and three were killed.

The most recent manifestation is the modern circus in which animal trainers 'defy death' in cages containing performing big cats, bears and other dangerous animals. The animal trainer becomes a surrogate gladiator, confronting man-eaters in the arena on behalf of an audience that wants to be frightened – at least, at a distance.

There are sometimes accidents. In January 1998, a British animal trainer was attacked by one of his performing tigers when the circus was visiting St Petersburg, Florida. During a press photo-call he was 'chuffing' – a friendly tiger greeting – with one tiger, when the one behind pounced and grabbed him by the head. The tiger was one the trainer had brought up from a cub. Circus workers and the trainer's brother fired fire extinguishers at the tiger and it let go. The young man was removed from the cage and the tiger was shot. The tiger had crushed his skull, but the man survived.

In February 1998, a circus accident at a winter quarters near Oxford saw a young man lose his arm to a tiger. He was closing a partition after feeding time when the tiger lunged and grabbed his arm, ripping it off and eating it. And in South Africa in April 1996, an Australian soap opera star from the show *Neighbours*, on a wildlife calendar photography shoot, was attacked from behind by one of the lions with which the young man had been posing. He had deep bites to the neck and long scratches down his back.

Zoos and drive-in wildlife parks have also had their share of 'man-eaters in action'. In August 1986, for instance, a group of children broke into the Camperdown Park wildlife centre after it had closed for the night, and a 10-year-old boy, who was thought to have been taunting a captive brown bear with a stick, was savaged. The bear ripped off the boy's arm just below the elbow and ate it. In September 1994, London Zoo launched a security review after a man climbed into the Asian lion's enclosure and was mauled by three lions. It was a re-run of an incident in December 1992, when a schizophrenic boy leaped into the same compound.

In March 1997, an animal handler at Savage Kingdom, Florida, lost a leg to a Siberian tiger he had been feeding. The animal was shot. In Israel, a young helper was savaged on the arms and legs and nearly killed by a spotted hyena in a wildlife sanctuary. There was no doubt that it was intent on feeding. It took chunks from the girl's arm and swallowed them. And, in December 1998, two escaped jaguars attacked a four-year-old boy in a wildlife park at Doué-la-Fontaine in Central France. The boy, together with his brother and sisters and some friends, were running ahead of their parents and heading towards a cage containing two cats. Unknown to them, the jaguars had tunnelled under the fence and were lying in wait. They pounced on the first child and killed him. One jaguar was shot by police marksmen, and the other captured and put down.

In 1998, a tiger adopted by animal lovers in Britain ripped off the arm of a five-year-old boy in Chachoengsao, Thailand. In the same year, a rare white Bengal tiger was being walked on a leash between its night kennel and day kennel at Newberry, Florida, when it was startled by the noise of construction workers nearby and attacked its trainer. The animal bit him on the neck and he died later in hospital. Six weeks later the same tiger attacked and killed another person, its co-owner, as she hand-fed it chicken necks. She was also bitten on the neck.

In January 1999, Thailand was in the news again when four tame but hungry tigers attacked their keeper at the Baan Sua (Tiger House) restaurant-cum-zoo in the Phrae district of northern Thailand. Poor business meant the tigers had not fed for several days so they took what food was available and hijacked the keeper when he entered their cage. In February 1999, a 37-year-old female keeper at the Fort Wayne Children's Zoo, Indiana was badly mauled, suffering serious scratches and cuts to her neck, back and chest. Against zoo policy, she had entered the cage to clean it while the tiger was still there. Another zoo-keeper nearby heard the woman's screams and came running. She directed a pepper spray at the tiger and, while it drew back, she was able to pull the injured keeper out.

A strange encounter, however, took place in rural England when, in June 1999, a farmworker at Armthorpe, Doncaster claimed he was attacked by a fully grown tiger. He was working on a fork-lift truck when the animal rushed out of the bushes and hissed at him, but did not attack. Local police say the incident was unlikely to have been a hoax as the man appeared genuinely terrified.

It is at times like these that people are reminded dramatically that the placid-looking creatures on the other side of the bars or outside the car are, in reality, wild and dangerous animals quite capable of indulging in a little man-eating. This occurred on two occasions at a safari park in the Gunma Prefecture, north-west of Tokyo. In November 1983, a tiger managed to get inside a tour bus and maul the tour guide, and in August 1997 another killed two people who got out of their car. Despite clear warnings of the danger, an elderly couple left their car and tried to carry a crying grandchild to its mother's car to be comforted. In full view of the other children, the couple were stalked and attacked by a Bengal tiger. They were both rushed to hospital with severe bite wounds all over their bodies, but later died. The grandchild was safe.

## CRYPTO-BEASTS

If predators – including those that you might come near in a zoo or wildlife park – are not frightening enough in themselves, why not invent some that are even more powerful and dangerous? Here, fact blurs with fiction, and it can be a devil of a job distinguishing one from the other. The science that attempts to do so is crypto-zoology, the study of unknown animals.

Sea serpents and lake monsters generally have not been known to take human victims, but there are a few exceptions. Lake Menbu on the Tibetan Plateau, for example, is host to a long-necked lake monster with a big head and 'body the size of a house'. It grabbed a farmer rowing across the lake on a raft and ate him, and also dragged a cow, tied up nearby, into the water.

The bunyip, an Australian crypto-beast with a dog-like head and piggy ears, was supposed to have been responsible for the disappearance of two stockmen at Lismore Lagoon in 1900.

If you do not believe in monstrous animals, how about a man-eating tree instead? In several books, seriously scientific authors have made claims for plants that would be more at home in *The Day of the Triffids* than in scientific journals. The ya-te-veo, for example, is a plant from South Africa and Central America that is supposed to eat not only insects like any other self-respecting carnivorous plant, but also larger living things, such as people! It was described by J. W. Buel in *Sea and Land*, published in 1887. Buel describes 'an acquaintance' telling him of a tree with a short, thick trunk and a crown of branches armed with barbs. The branches do not wave about in the breeze like other trees, but drape on the ground in a circle radiating from the trunk. Should anyone step into the ring, he or she is enveloped in the spiny branches and drawn up towards the centre. The barbs pierce the body and the branches squeeze until every drop of blood is absorbed by the plant.

An edition of *The Illustrated London News*, dated 27 August 1892, has a report from Dr Andrew Wilson of man-eating vegetation living in swamps on the edge of Lake Nicaragua. He heard the story from a naturalist by the name of Mr Dunstan who was examining the plants of the swamp. Suddenly Dunstan heard his dog yelp in pain, and running to the spot found the poor animal enveloped in a network of rope-like tissues, much like a weeping willow without leaves. The branches exuded a thick gum from pores. Taking a knife, Dunstan tried to cut the dog free, and as he did so, the branches coiled around his hand. When he finally extricated both the dog and his hands, he was shocked to see his arms were blistered and red, and the dog's fur was bloodstained. Dunstan wrote that the plant was well known to the local villagers. They would throw it a large lump of meat which it would suck dry, like a spider with a fly; at least, that was the local folk tale.

Wilson, give him his due, suspected the story to be pure fiction and invited his readers to write in about similar discoveries. He thought they might be able 'to inform me whether or not the matter is a "plant", vulgarly speaking, in another sense'. It must have come as a bit of a surprise, then, when he received another report of a carnivorous plant, this time from the Sierra Madre in Mexico. In an edition dated 24 September 1892, he described a tree with slimy, snake-like branches that trapped birds. It ingested the blood from suckers on the branches, according to the traveller who wrote to Wilson, and would grab a person's arm if he got too close.

Believe them or not, the stories feed that inner fear we have of the dark and eerie forest where branches like clawing hands are ready to grab anyone who should walk by. The greatest number of man-eating tree stories, however, emanate not from Central and South America, but from Madagascar which has been known since time immemorial as 'the land of the man-eating tree'. This was something that struck Chase Salmon Osborn as worthy of investigation and he travelled the length and breadth of the island in search of the tree. Locals knew of it but nobody would show him it, and so when he came to write about his adventures in *Madagascar: Land of the Man-eating Tree*, published in 1924, he had to resort to unsubstantiated stories from earlier travellers.

In 1878, a German explorer, Carle Liche, is said to have written to a Polish colleague about an incident that took place involving a primitive tribe of pygmies who grew no more than 56in (1.42m) in height, and lived in limestone caves. Out walking with the tribe one day, Liche and his companion came across a strange tree. Its trunk was like an 8ft (2.4m) high pineapple, and its eight leaves radiated from the crown like those of the agave. At the apex of the trunk was a receptacle containing a clear treacly liquid, like the centre of a bromeliad. Stretching from the tree were also 'long, hairy green tendrils' that stretched out for 8ft (2.4m) or more. Around the cup six 'white almost transparent palpi', each about 6ft (1.8m) tall, 'reared themselves towards the sky, twirling and twisting with marvellous incessant motion'.

By the time Liche and his companion had taken all this in, the pygmies had become very excited and began to behave strangely. A woman was isolated from the crowd, and at spear-point she was forced to climb into the tree and urged to drink the liquid in the crown. In an instant, she seemed possessed 'with wild frenzy in her face', according to Liche, and the tree itself suddenly sprang into life! The palpi wrapped around her like snakes and, while she alternately screamed and convulsed with demonic laughter, the tendrils were gradually drawn in and tightened around her like a constricting snake squeezing its victim. The thick leaves, meanwhile, closed and pressed down on the woman like a hydraulic press. The viscid liquid oozed from the crown and, mixed with the blood and visceral fluids from the human sacrifice, it flowed down the trunk. The rest of the tribe ran to collect and drink it, and they became intoxicated to such an extent they indulged in what Liche described as 'a grotesque and indescribably hideous orgy'. A few days later, Liche re-visited the tree. The leaves and tendrils had unfurled and all that was left of the woman was a clean, white skull . . . or so the story goes.

Crypto-zoologist and biochemist Roy Mackal, of the University of Chicago, tried to get to the bottom of the affair and was confronted by a tangled web of popular newspaper stories and reports in semi-scientific journals. In his *Searching for Hidden Animals*, published in 1980, he wrote that at least one missionary report confirmed that legends did exist about man-eating trees, but that they were just that – legends.

Fellow of the Royal Zoological Society, the Reverend George Shaw, published an account – *Madagascar and France* – of Madagascar's people and resources in the 1880s, and concluded that the tales of the man-eating tree were 'in the imagination of the writers'. But, man-eating trees were not to stop there. Mackal discovered more.

The legend of the man-eating tree can be heard in many of the lands bordering the Indian Ocean, including India where in a poem entitled 'Story' by Lalla-ji, the priest tells of a tree without blossom to which fertiliser was added to make it bloom. The fertiliser, however, was a human child buried alive amongst the roots!

South Pacific islands also have their predatory plants. In *Myths and Legends of Flowers, Trees, Fruits and Plants*, Charles Skinner writes of a Captain Arkright who was sailing the islands in 1581. Arkright came across a local story about El Banoor – the Island of Death – on which grew a frightening flower. The flower, whose description resembles a giant *Rafflesia* (a parasite on a tropical vine and the largest flower in the world), lured its victims with the aid of a strong and soporific odour to come and rest among its petals. They would then fold over and dissolve the body with acid released from the calyx.

Alas, the prospect of a man-eating tree, though a likely box-office hit at Kew, is an improbability. The largest known carnivorous plant is a pitcher plant (*Nepenthes rajah*) from Borneo, which can have a pitcher containing 7 pints (4 litres) of liquid and possessing an opening 1ft (0.3m) across. Animals as big as birds and rodents might drop in, but not people!

# REFERENCES

**WHY MAN-EATERS?**

Anon (1999) Farmworker tells of tiger attack terror. *Eastern Daily Press*, 18 June 1999.

Armstrong, J., Sawyer, P. and Rock, L. (1998) Oxford tiger attack. *Daily Record, Evening Standard*, 26 February 1998, and *The Mirror*, 27 February 1998.

Greenfield, Joanna (1997) I was eaten by a hyena. *Sunday Telegraph*, 16 November 1997.

Guggisberg, C. A. W. (1975) *Wild Cats of the World*. David and Charles.

Hoyer, Meghan (1999) Tiger attacks zookeeper. *Journal Gazette*, (Fort Wayne), 28 February 1999.

Poli, François (1959) *Sharks Are Caught at Night*, Henry Regnery.

Walker, A. et al (1998) *Journal of the American Chemical Society*, 121:128.

**WOLVES AND WILD DOGS**

Anon (1995) Man bites wolf in attack. Associated Press, 15 October 1995.

Anon (1995) Wolf attacks assume dangerous scale in Krygyzstan. BBC Monitoring Service, 12 January 1999.

Begley, Sharron et al (1991) Return of the Wolf. *Newsweek*, 12 August 1991, p. 44–50.

Caras, Roger (1964) *Dangerous to Man*. Barrie and Jenkins Ltd.

Ford, Correen (1999) Dog had toddlers head in its jaws. *The Journal*, 1 April 1999.

Goss, Richard (1986) *Maberley's Mammals of Southern Africa*. Delta Books.

Harries, Kate (1998) Fearless grey wolf attacks toddler in Algonquin Park. *The Toronto Star*, 29 September 1998.

Harrington, Fred, Paquet, Paul (1982) *Wolves of the World*. Noyes Publications.

Klinghammer, Erich (1996) Captive Non-Human scoialised wolves kill caretaker in a Canadian Forest and Wildlife Reserve. *Wolf Magazine*, 5 May 1996.

McGuigan, Ciaran (1999) Girl is bitten by 'pet' wolf. *Belfast Telegraph*, 29 April 1999.

Mech, David L. (1970) *The Wolf*. Constable and Co. Ltd.

Naylor, Janet, Ackerman, Jane (1997) Dog attacks are up in state and nation – so are lawsuits. *The Detroit News*, 12 August 1997.

O'Neill, Sean (1999) 136 stitches for boy attacked by Rottweiler. *The Daily Telegraph*, 4 August 1999.

Riddle, Maxwell (1979) *The Wild Dogs in Life and Legend*. Howell Book House Inc.

Shahi, S. P. (1982) Status of the grey wolf in India – a preliminary survey. *Journal of the Bombay Natural History Society*, 79:493–502.

Scott, P. A., Bentley, C. V., Warren, J. J. (1985) Aggressive Behaviour by Wolves Towards Humans. *J. Mamm.*, 66 (4): 807–809.

Zimen, Erik (1981) The Wolf: his Place in the Natural World. Souvenir Press.

## MAN-EATING TIGERS AND TIGERS OF THE SWAMP

Abdul, Jasmibin (1998) The distribution and management of the Malayan tiger in Peninsular Malaysia. Year of the Tiger Conference Abstracts.

Anon (1972) Man-eating Tiger Problem (report on Hendrichs' WWF survey) *Orynx* 11 (4): 231.

Anon (1997) Tiger attacks in Sumatra block reforestation project. Deutsche Press-Agentur, 23/26 June 1997.

Buckland, C. T. (1889) A black tiger. *Bombay Natural History Society*, vol. 4, pp.149–150.

Choudhury, M. K., Sanyal,P. Use of eletroconvulsive shocks to control tiger predation on human beings in Sundarbans Tiger Reserve. *Tigerpaper*?? pp.1–5.

Jackson, P. (1983) Tigers and Men. WWF Monthly Report, April 1983, Project 1000, Operation TYiger, pp.487–491.

Jackson, P. (1984) *Shocking Tiger Tales*. Earthscan Feature.

Jackson, P. (1985) Man-eaters. *International Wildlife*. Nov/Dec 1985, pp.4–11.

Jackson, P. (1985) Deaths as a result of tiger attacks. *Cat News* 2, Spring 1985.

http://lynx.uio.no/catfolk/cnissues/cn02-14.htm

Jackson, P. (1985) Tiger kills British ornithologist in India. *Cat News* 3, Summer 1985.

http://lynx.uio.no/catfolk/cnissues/cn03-10.htm

Jackson, P. (1985) Tiger problems in Nepal. *Cat News* 3, Summer 1985.

http://lynx.uio.no/catfolk/cnissues/cn03-11.htm

Jackson, P. (1985) Tiger problems in the USSR. *Cat News* 3, Summer 1985.

http://lynx.uio.no/catfolk/cnissues/cn03-12.htm

Jackson, P. (1986) Man-eater shot in Nepal. *Cat News* 5, Summer 1986.

http://lynx.uio.no/catfolk/cnissues/cn05-15.htm

Jackson, P. (1987) Girl survives tiger attack in Nepal. *Cat News* 7, Summer 1987.

http://lynx.uio.no/catfolk/cnissues/cn07-13.htm

Jackson, P. (1988) Tiger/Human conflict around Dudhwa National Park. *Cat News* 8, Autumn 1988.

http://lynx.uio.no/catfolk/cnissues/cn08-03.htm

Jackson, P. (1989) Man-eaters and masks. *Cat News* 11, Summer 1989.

http://lynx.uio.no/catfolk/cnissues/cn11-11.htm

Jackson, P. (1989) Tiger attacks in Bangladesh. *Cat News* 11, Summer 1989.

http://lynx.uio.no/catfolk/cnissues/cn11-12.htm

Jackson, P. (1989) Tiger attacks around Chitwan National Park. *Cat News* 11, Summer 1989.
http://lynx.uio.no/catfolk/cnissues/cn11-13.htm
Jackson, P. (1989/90) Hungry Sundarbans Tigers. *Cat News* 12, Winter 1989/90.
http://lynx.uio.no/catfolk/cnissues/cn12-11.htm
Kholenko, Victor (1998) Hunters kill man-eating tiger. *The Vladivostok*, issue 158.
Macgowan, T. (1998) Man-eater in Chitwan. The Tiger Information Center. http://www.5tigers.org/chitwan.htm
Mills, Stephen (1992) Stars in Stripes. *BBC Wildlife* 10 (11) pp.32–42.
Mountford, Guy (1973) *Tigers*. Wildlife International series. David and Charles.
Parihar, A. S. The man-eater of Papra.
Rishi, Vinod (1988) Man, Mask and Man-eater. *Tiger Paper*, July-September 1988, pp.9–14.
Sankhala, Kailash (1978) *Tiger: the Story of the Indian Tiger*. Collins.
Singh, Arjan (1984) *Tiger! Tiger!* Jonathon Cape.
Singh, K. M. (1997) Marksmen hunt child-eating tiger. *Daily Telegraph*, 25 January 1997.
Sunquist, Fiona and Sunquist, Mel (1988) *Tiger Moon*. University of Chicago Press.

**MAN-EATING LIONS**
Blair, David (1999) Briton left tent before lion attack. *Daily Telegraph*, 9 August 1999.
Brown, Thomas (1905) *Interesting Anecdotes of the Animal Kingdom*. London.
Jackson, P. (1987) Man-eating lions in Tanzania. *Cat News* 6/7, Spring and Summer 1987.
http://lynx.uio.no/catfolk/cnissues/cn06-06.htm **and** http://lynx.uio.no/catfolk/cnissues/cn07-19.htm
Jackson, P. (1989/90) Man-eating lions killed. *Cat news* 12, Winter 1989/90. http://lynx.uio.no/catfolk/cnissues/cn12-15.htm
Munnion, Christopher (1998) Big cats get a taste for illegal immigrants. *Daily Telegraph*, 26 August 1998.

Patterson, J. H. (1907) The Man-eaters of Tsavo. Macmillan, London.

Vosper, Robert (1998) The Man-eater of Mfuwe. *In the Field* (membership publication of the Field Museum), November/ December 1998.

## CATS: SILENT HUNTERS

Anon (1994) Mountain lion increases; so do human conflicts. *National Geographic*, December 1994, p.146.

Anon (1998) Mountain lion killed after charging group. *Boston Globe*, 1 January 1998.

Anon (1998) Mother fights off lion. *Evening Standard*, 9 June 1998.

Brander, A. A. Dunbar (1927) *Wild Animals in Central India*. London.

Brown, Thomas (1905) *Interesting Anecdotes of the Animal Kingdom*. London.

Corbert, Jim (1947) *The Man-eating Leopard of Rudraprayag*. India.

Hitchens, William (1937) *Africa's Mystery Beasts*. Discovery, London.

Jackson, P. (1987) Man-eating leopard in Nepal. *Cat News* 6, Spring 1987. http://lynx.uio.no/catfolk/cnissues/cn06-05.htm

Jackson, P. (1987) Man-eating leopard shot in Nepal. *Cat News* 7, Summer 1987. http://lynx.uio.no/catfolk/cnissues/cn07-08.htm

Jackson, P. (1988) Cougar scares woman in restroom. *Cat News* 8, Autumn 1988. http://lynx.uio.no/catfolk/cnissues/cn08-20.htm

Jackson, P. (1989) Leopard attacks in Nepal. *Cat News* 9, Spring 1989. http://lynx.uio.no/catfolk/cnissues/cn09-11.htm

Jackson, P. (1989) Leopard stoned to death. *Cat News* 11, Summer 1989. http://lynx.uio.no/catfolk/cnissues/cn11-04.htm

Kala, Arvind (1998 Killers on the Loose. *Asiaweek*, 24 April 1998.

La Guardia, Anton (1998) Leopard killed with a screwdriver. *Daily Telegraph*, 2 October 1998.

Perry, Richard (1970) *The World of the Jaguar*. David and Charles.

Toit, Julienne du (1998) Man on the menu. *BBC Wildlife*, December 1998, p.25.

Zaidle, Don (1997) *American Man-Killers*. Safari Press.

## THE THREE BEARS

Anon (1998) Finns hunt for bear after killing. *The Independent*, 19 June 1998.

Anon (1983) Grizzly bear killed after eating sleeping camper. *Guardian*, 28 June 1984.

Anon (1998) Woman dies in attack by bear. *Evening Standard*, 12 October 1998.

Anon (1999) Man killed by bear in Hokkaido. *Daily Yomiuri*, 9 May 1999.

Anon (1999) Swan halts bear attack. *Daily Record*, 17 June 1999.

Anon (1999) Polar bear attack (Hudson Bay). *The Journal*, 17 July 1999.

Anon (1999) Bear attack (Tomahawk Bay). *Daily Record*, 11 August 1999.

Anon (1999) Polar bear attack (Svarlbad). *Washington Post*, 25 August 1999.

Bromley et al (1992) *Safety in Bear Country*. Department of Renewable Resources Safety in Bear Country Program, Government of the Northwest Territories.

Cramond, Mike (1981) *Killer Bears*. Outdoor Life Books, Charles Scribner's Sons.

Herrero, Stephen (1970) Human injury inflicted by grizzly bears. *Science* 170: 593–598.

Herrero, Stephen (1976) Conflcits between man and grizzly bears in the national parks of North America. In Third International conference on bear research and management, eds. Pelton, M., Lentfer, J. W., Folk, G. E. IUCN New Series 40, pp.121–145.

Herrero, Stephen (1985) *Bear Attacks: Their Causes and Avoidance*. Lyons and Burford.

Kaniut, Larry (1983) *Alaska Bear Tales*. Alaska Northwest Books.

Lightfoot, Rebecca (1998) Fear the Ferret. *Country Life*, March 1998, p.182.

Olsen, Jack (1969) *Night of the Grizzlies*. Signet Books.

Samstag, Tony and Peter Markham (1998) Arctic Britons forced to kill polar bears. *Evening Standard*, 11/12 August 1998.

Scobie, William (1984) Killer grizzly bears stoned on drugs. *Observer*, 26 August 1984.

Struzik, Ed (1987) Nanook: in the tracks of the great wanderer. *Equinox* 31, January/February 1987, pp.18–32.

Tilson, Ronald (1983) Carcass Protocol. *Natural History*, March 1983, pp.42–47.

Whitlock. S. C. (1950) The black bear as a predator of man. *J. Mammal*. 31 (2): 135–138.

## HYENAS, PIGS AND EAGLES

Kingdom, Jonathan (1977) East African mammals: an atlas of evolution in Africa, vol. III, Part A (Carnivores). Academic Press.

Lightfoot, Rebecca (1998) Fear the Ferret. *Country Life*, March 1998, p.182.

Mackal, Roy (1980) *Searching for Hidden Animals*. Doubleday, New York.

Michel, John and Robert, Richard (1982) *Living Wonders*. Thames and Hudson.

Williams, Philip (1989) Man-eating hyenas terrify tribesmen. *Sunday Times*, 25 June 1989.

## MAN-EATING CROCODILES

Anon (1994) Saurians ravage Madagascar villages. ROI Madagascar, December 1994.

Anon (1995) Probem crocodile. *Kenya Times*, 25 January 1995.

Anon (1995) Terror crocodile killed by wardens. *Daily Nation* (Kenya), 28 February 1995.

Anon (1995) Recent crocodile attacks in PNG. *Post Courier*, March/May 1995.

Anon (1995) Part human, part crocodile. *Discover* 16 (5).

Anon (1995) Crocodiles dressed like humans burned alive. *The Orlando Sentinel*, 25 September 1995.

Anon (1996) Crocodiles Turn Back Nile Paddlers. Notes from the Nile, 6 February 1996. http://www.adventureonline.com

Anon (1996) Florida alligators feeling cramped. *Aiken Standard*, 30 August 1996.

Anon (1998) Schoolgirl survives attack by crocodile. *Sunday Mirror* and *Mail on Sunday*, 8 February 1998.

Anon (1999) Cairns schoolgirl survives attack by crocodile. *Sunday Mirror*, 8 February 1999 and *Daily Mirror*, 10 February 1999.

Behra, Olivier (1995) Reports of crocodile attacks on people in Madagascar. Crocodile Specialist Group Newsletter.

Lazcano-Berrero, Marco A. (1996) Crocodile attacks in Cancun. Crocodile Specialist Group Newsletter 154D.

Neumann, Arthur (1898) *Elephant-Hunting in East Equatorial Africa.* London.

Pinney, Peter (1976) *To Catch a Crocodile.* Angus and Robertson.

Pooley, Tony (1982) *Discoveries of a Crocodile Man.* Collins.

Ranot, Shlomi (1996) Survey of Tanzanian crocodiles and human conflicts. Crocodile Specialist Group Newsletter.

Soorae, P. S. (1994) Nile crocodiles in Somalia. Crocodile Specialist Group Newsletter.

Steele, Rodney (1989) *Crocodiles.* Christopher Helm.

Vyas, R. (1993) Recent cases of man-eating by the mugger in Gujurat State. *Hamadryad* (Madras), 18:48–49.

Webb, G. (1994) Vagrant crocodile. Crocodile Specialist Group Newsletter.

## DRAGONS AND SERPENTS

Anon (1998) 24ft snake found with man in its stomach. *The People*, 25 October 1998.

Diamond, Jared (1987) Komodo dragons and pygmy elephants. *Nature* 326, p.832.

Diamond, Jared (1992) The evolution of dragons. *Discover*, December 1992, pp.72–80.

Letts, Quentin (1996) Hungry python kills owner. *The Times*, 11 October 1996.

Mitchell, P. B. (1987) Here be Komodo dragons (correspondence). *Nature* 329, p.111.

Neumann, Arthur (1898) *Elephant-Hunting in East Equatorial Africa.* London.

Smith, Kevin (1996) Writhing in agony. *Daily Mail*, 23 August 1996.

Stevens, Jane Facing the Dragons. *International Wildlife*, pp.30–35.

**SHARK ATTACKS**
Anon (1998) Shark attack (Loch Lomond). *Daily Mirror*, 26 January 1998.
Anon (1998) Recife shark attack. *Daily Mirror*, 5 May 1999.
Anon (1998) Pair in shark attack. *Daily Record*, 3 August 1998.
Anon (1998) Vero Beach attack. *Washington Post, New York Yimes, The Journal, Western Morning News, Daily Mail, Birmingham Post, The Mirror, Evening Standard*, 25 November 1998, and *Independent*, 9 December 1998.
Anon (1998) Increase in shark attacks. *Sunday Times*, 6 December 1998.
Anon (1999) Hawaii kayak attack. *Daily Record*, 24/25 March 1999, *Daily Express*, 25 March 1999.
Anon (1999) Shark bites Big Island man. *The Honolulu Star-Bulletin*, 22 July 1999.
Heimersson, Mats (1999) Greenland Sharks. *National Geographic* (Forum), January 1999.
Morgan, Gary (1998) Surf man spat out by shark. *The Mirror*, 24 April 1998.
Thomas, Pete (1999) A tale of horror in Hawaiian waters. *Los Angeles Times*, 27 March 1999.
Thorne, Frank (1998) Fisherman dies after being savaged by Jaws-type shark. *Evening Standard*, 29 June 1998.
Wallett, Tim (1978) *Shark Attack and Treatment of Victims in Southern African Waters*. Purnell.

**WHITE DEATH**
Anon (1997) Great white attack on diver and underwater scooter. *Scottish Daily Record, Sunday Mail*, and *The Mirror*, 12 November 1997.
Anon (1998) Surfer gives great white a great fight. *Daily Mail*, 1 June 1998.
Anon (1998) Pair in shark attack. *Daily Record*, 3 August 1998.
Anon (1999) Surfer in shark attack. Associated Newspapers, 25 February 1999.
Klimley, Peter A. (1994) The predatory behaviour of the white

shark. *American Scientist* 82:122–133.

Morgan, Gary (1998) Surf man spat out by shark. *The Mirror*, 24 April 1998.

Munnion, Christopher (1998) Surfers fall prey to great white shark. *Sunday Telegraph*, 5 July 1998.

Thorne, Frank (1998) fisherman dies after being savaged by Jaws-type shark. *Evening Standard*, 29 June 1998.

Wallett, Tim (1978) *Shark Attack and Treatment of Victims on Southern African Waters*. Purnell.


## WATER-BEASTS

Anon (1997) Beaches re-open after Canadian boy gashed by ocean predator. *Boston Globe*, 26 December 1997.

Anon (1999) Killer whale tries to bite trainer. Associated Press, 14 June 1999.

Doak, Wade (1999) The Friendly Killers. *Diver*, August 1999.

French, T. W. (1981) Fish attack on black guillemot and common eider in Maine. *Wilson Bull*, 93 (2) 279–280.

Hoyt, Erich (1981) *Orca: the Whale Called Killer*. Dutton.

Hoyt, Erich (1992) *The Performing Orca*. Whale and dolphin conservation Society.

Sazima, I., de Andrade-Guimaraes, S. (1987) Scavenging on human corpses as a source for stories about man-eating piranhas. *Environ. Biol. Fish.* 20 (1):75–77.

Spaeth, Anthony (1992) *Healing the Holy Ganges*. Tomorrow.

Spong, Paul (1974) *Mind in the Waters*, ed. McIntyre, J. Scribner.

Taylor, Kizzy (1999) Monster fish leaps to grab tourist in boat. *Daily Mail*, 10 February 1999.

Thomas, Pete (1998) Bodega's big squid. *Los Angeles Times*, 6 November 1998.

Zhal, Paul (1970) Seeking the truth about the feared piranha. *National Geographic*, November 1970, pp.714–733.


## MINI-BEASTS AND BLOODSUCKERS

Delpietro, H. A. (1989) Case reports on defensive behaviour in equine and bovine subjects in response to vocalization of the common vampire bat. *Appl. Anim. Behav. Sci.* 22 (3–4):377–380.

Homewood, Brian (1994) Vampire fish show their teeth. *New Scientist*, 3 December 1984, p.7.

Kuerton, L., Schmidt, U., Schaefer, K. (1984) Warm and cold receptors in the nose of the vampire bat. *Naturwissenschaften* 71 (6):327–328.

Mackey, Mary *I Should Have Stayed at Home*, ed. Repoport and Casterera. Book Passage Press, Berkley, California.

## CANNIBALS, SACRIFICES AND MAN-EATING TREES?

Arens, William (1997) Man is off the menu. *Times* Higher Education Supplement, 12 December 1997.

Bakewell, Sarah (1999) Cooking with Mummy. *Fortean Times* 122.

Beck, B. F., Smedley, D. (1947) *Honey and your Health*. Museum Press, London.

Croll, O. (1909) *Basilica Chymica*. Frankfurt.

Gibbons, Ann Archaeologists rediscover cannibals. *Science* 277:635.

Goodall, Jane (1990) *Through a Window*. Houghlin Mifflin.

Gordon-Brube, K. (1998) Anthropophagy in post-Renaissance Europe. *American Anthropologist* 90:405.

Hearn, Lafcadio (1882) A strange tale of cannibalism. *The Times-Democrat*, 15 October 1882.

Le Fevre, N. (1664) *A Compleat Book of Chymistry*, part 1 (translated by P. de Cardonel). London.

McKie, Robin (1998) The People-eaters. *New Scientist*, 14 march 1998.

Metraux, Alfred (1957) *Easter Island: A Stone-Age Civilization of the Pacific*. Oxford University press.

Pfenig, D. W., Ho, S. G., Hoffman, E. A. (1998) Pathogen transmission as a selective force against cannibalism. *Animal Behaviour* 55:1255–1261.

Pomet, P.(1712) *A Compleat History of Drugs*. London.

Reed, B. E. (1931) Chinese Materia Medica, v:6 Animal Drugs. *Peking Natural History Bulletin*.

Rykovtseva, Yelena (1996) Will cannibalism become an epidemic? *Moskovskiye Novosti*, 25 August 1996.

Tannahill, Reay (1996) *Flesh and Blood*. Abacus, London.

Turner, Jacqueline, Man Corn, Christy, *Cannibalism and Violence in the Prehistoric American Southwest*. University of Utah Press.

Walker, A. et al (1998) *Journal of the American Chemical Society* 121:128.
White, Time (1992) *Prehistoric Cannibalism at Mancos.* Princeton.

# INDEX